CISM Courses and Lectures

The series presents lecture notes, monographs, edited works and proceedings in the field of Mechanics, Engineering, Computer Science and Applied Mathematics.
Purpose of the series is to make known in the international scientific and technical community results obtained in some of the activities organized by CISM, the International Centre for Mechanical Sciences.

International Centre for Mechanical Sciences

Courses and Lectures Vol. 554

For further volumes:
www.springer.com/series/76

Francisco Chinesta · Pierre Ladevèze
Editors

Separated Representations and PGD-Based Model Reduction

Fundamentals and Applications

Springer

Editors

Francisco Chinesta
Ecole Centrale de Nantes, France

Pierre Ladevèze
Ecole Normale Supérieure de Cachan, France

ISSN 0254-1971
ISBN 978-3-7091-1793-4 ISBN 978-3-7091-1794-1 (eBook)
DOI 10.1007/ 978-3-7091-1794-1
Springer Wien Heidelberg New York Dordrecht London

All contributions have been typeset by the authors

Printed on acid-free paper

Springer is part of Springer Science+Business Media (www.springer.com)

PREFACE

Today, in spite of the impressive progress which has been achieved in the past decade in modeling, numerical analysis, discretization techniques and computer science, many science and engineering problems remain inextricable because their numerical complexity or the restrictions imposed by the operational requirements (for example, real-time computing on deployed platforms) make it too expensive to solve them using today's technologies. These include a group of problems with no apparent relation to one another, but which can be treated in a unified manner, as this course will show. What these problems have in common is our inability or our insufficient knowledge to solve them numerically in a direct, traditional way. Thus, in order to get a solution, it is necessary to resort to some sort of model order reduction.

We begin this course popular by describing the construction of reduced models through a review of Proper Orthogonal Decomposition (POD) and reduced-basis models, including their mathematical foundations and some challenging applications. Then, we move to a new generation of simulation strategies based on the use of separated representations (space-parameters, space-time, space-time-parameters, space-space, ...), which have led to what is known as Proper Generalized Decomposition (PGD) techniques. Because such representations enable one to do away with dimensionality restrictions, the models can be enriched by treating parameters as additional coordinates, leading to fast and inexpensive online calculations based on richer offline parametric solutions. Separated representations are analyzed in detail in the course, from their mathematical foundations to their most spectacular applications. We illustrate how such an approximation could evolve into a new paradigm in computational science, enabling one to circumvent various computational issues in a vast array of applications in engineering science.

The course is intended for doctoral students, junior or senior researchers and practicing engineers working in the area of simulation software, who are confronted with the severe limitations of standard simulation techniques in the resolution of complex models.

We would like to thank our esteemed colleagues and friends Elias Cueto, Antonio Huerta, Boris Khoromskij and Gianliugi Rozza for accepting our invitation to lecture in our course and contributing very significantly to the lecture notes (except for B. Khoromskij due to health problems).

We would also like to express our thanks to CISM, especially to the Rectors and staff, for their help and cooperation in organizing the course and printing the lecture notes.

Francisco Chinesta and Pierre Ladeveze

CONTENTS

Model Order Reduction based on Proper Orthogonal Decomposition

Elías Cueto*, Francisco Chinesta† and Antonio Huerta‡
* I3A, Universidad de Zaragoza
Maria de Luna s/n, 50018 Zaragoza, Spain
† ESI Group Chair on Advanced Computational Manufacturing Processes
GEM, UMR CNRS - Centrale Nantes
1 rue de la Noe, BP 92101, F-44321 Nantes cedex 3, France
‡ Laboratori de Calcul Numeric
Universidad Politecnica de Cataluña
Jordi Girona 1, Campus Nord, C2, E-08034 Barcelona, Spain

Abstract In this chapter we review the basics of classical model order reduction techniques based on Proper Orthogonal Decomposition (POD), Principal Component Analysis (PCA) or Karhunen-Loève decompositions.

1 Introduction

Simulation has become essential in nowadays industrial processes in general, and in forming processes in particular. The appearance of the finite element method in the 50's, and the continuing growth of computer performance have allowed the industry to simulate processes and products. But today the problem remains precisely in the complexity of state-of-the-art computer simulations. These are surprisingly accurate and complex, being able to simulate complex multi-physic, multi-scale phenomena with a high degree of fidelity. Industry faces today a different challenge: to have these simulation results in a short time so as to be able to take decisions without delays. Typically, high-tech industries such as aeronautic manufacturers accept simulations able to run in one night. This makes it possible to take decisions in the morning of the following day. However, many state-of-the-art simulations take today days or even weeks.

Model order reduction (MOR) or, simply, model reduction techniques exist many decades ago. These techniques allow for a simplified representation of the evolution of physical systems, employing a very reduced number of degrees of freedom. Based on the above considerations, MOR techniques have gained a tremendous popularity in the last years due precisely to their

F. Chinesta, P. Ladevèze (Eds.), *Separated Representations and PGD-Based Model Reduction Fundamentals and Applications*, CISM International Centre for Mechanical Sciences
DOI 10.1007/ 978-3-7091-1794-1_1 © CISM Udine 2014

ability to greatly reduce computational cost at a price of only a minimal loss of accuracy, that can be nowadays fully quantified.

MOR techniques seem, therefore, to be an appealing choice for their generalization in industrial contexts, and particularly in the simulation of forming processes. Many forming processes involve complex multi-physic phenomena in a coupled manner, with inherent strong non-linearities, uncertainty in parameter determination, etc. Thus, simulation in this context involves very complex techniques that, despite their accuracy and sound theoretical basis, are strongly CPU-consuming.

In this first chapter of the book we deal with the basis of the most popular technique for model reduction: Proper Generalized Decomposition (or Karhunen-Lo?ve transform, or Principal Component Analysis), a technique discovered and re-discovered through time in virtually all scientific speciality. We deal with the basic ingredients as well as its main associated difficulties, those responsible of the birth of a new generation of techniques such as Reduced Basis or Proper Generalized Decomposition, for instance.

2 Model reduction: extracting relevant information

The basic ingredient of POD is the need to complete some full-order problems in order to extract from them "relevant" information to be used in "similar" problems. Both "relevant" and "similar" are here quoted, since their precise meaning will be clear afterwards. It is expected that we can evaluate some general patterns in the solution of the full problems so as to employ them as basis functions for subsequent modifications of the problem, and that this relevant information is low dimensional.

Assume now that, say, a finite element approximation is constructed on top of the problem at hand. Consider a mesh having M nodes. If the problem is transient, it will be necessary to compute at each time step M values (assuming that the solution is a scalar field). For non-linear problems, this implies the solution of at least one linear algebraic system of size M at each time step. It is easy to notice the complexity of the problem for large models.

Is it possible to determine of the solution lives in a space of much lower dimensions that that spanned by the finite element mesh? In principle yes. And POD provides an efficient means to do it.

2.1 An introduction to classical Proper Orthogonal Decomposition

Assume now that the solution field $u(\mathbf{x}, t)$ is known, for the already computed full-order solutions, at the nodes $\mathbf{x}_i$ of a spatial mesh for discrete time instants $t_m = m \cdot \Delta t$, with $i \in [1, \cdots, M]$ and $m \in [0, \cdots P]$. We employ the notation $u(\mathbf{x}_i, t_m) \equiv u^m(\mathbf{x}_i) \equiv u_i^m$ and define $\mathbf{u}^m$ as the vector of nodal values u_i^m at time t_m. POD provides an efficient means to obtain the most typical or characteristic structure $\phi(\mathbf{x})$ among these $u^m(\mathbf{x})$, $\forall m$ (Ryckelynck et al., 2006). To this end, we maximize the functional

$$\alpha = \frac{\sum_{m=1}^{P} \left[\sum_{i=1}^{M} \phi(\mathbf{x}_i) u^m(\mathbf{x}_i) \right]^2}{\sum_{i=1}^{M} (\phi(\mathbf{x}_i))^2}, \tag{1}$$

which can be shown equivalent to solve the following eigenvalue problem:

$$\mathbf{c}\boldsymbol{\phi} = \alpha \boldsymbol{\phi}. \tag{2}$$

Here, the vector $\boldsymbol{\phi}$ has as the i-th component $\phi(\mathbf{x}_i)$, and $\mathbf{c}$ is known as the two-point correlation matrix

$$c_{ij} = \sum_{m=1}^{P} u^m(\mathbf{x}_i) u^m(\mathbf{x}_j) = \sum_{m=1}^{P} \mathbf{u}^m \cdot (\mathbf{u}^m)^T, \tag{3}$$

which is symmetric and positive definite. With the matrix $\mathbf{Q}$ defined as

$$\mathbf{Q} = \begin{pmatrix} u_1^1 & u_1^2 & \cdots & u_1^P \\ u_2^1 & u_2^2 & \cdots & u_2^P \\ \vdots & \vdots & \ddots & \vdots \\ u_M^1 & u_M^2 & \cdots & u_M^P \end{pmatrix}, \tag{4}$$

we have

$$\mathbf{c} = \mathbf{Q} \cdot \mathbf{Q}^T. \tag{5}$$

2.2 Reduced-order modeling based on POD

The main objective being to obtain a reduced-order model, we solve Eq. (2) and keep the N most important eigenvectors $\boldsymbol{\phi}_i$, i.e., those associated with the eigenvalues belonging, say, to the interval defined by the highest eigenvalue α_1 and α_1 divided by a large enough number (e.g. 10^8). If POD works is just because N is very often (not always!) found to be much lower than M. These N eigenfunctions $\boldsymbol{\phi}_i$ are then used as basis functions

(Ritz-like) to approximate the sought function $u^m(\mathbf{x})$, $\forall m$. Let us define the matrix $\mathbf{B} = [\boldsymbol{\phi}_1 \cdots \boldsymbol{\phi}_N]$, i.e.

$$\mathbf{B} = \begin{pmatrix} \phi_1(\mathbf{x}_1) & \phi_2(\mathbf{x}_1) & \cdots & \phi_N(\mathbf{x}_1) \\ \phi_1(\mathbf{x}_2) & \phi_2(\mathbf{x}_2) & \cdots & \phi_N(\mathbf{x}_2) \\ \vdots & \vdots & \ddots & \vdots \\ \phi_1(\mathbf{x}_M) & \phi_2(\mathbf{x}_M) & \cdots & \phi_N(\mathbf{x}_M) \end{pmatrix}. \quad (6)$$

If, for instance, an explicit time-stepping scheme is used to compute the discrete solution $\mathbf{u}^{m+1}$ at time t^{m+1} a linear algebraic system like

$$\mathbf{G}^m \; \mathbf{u}^{m+1} \;=\; \mathbf{H}^m. \quad (7)$$

must be solved at each time step.

A reduced-order model is then obtained by projecting $\mathbf{u}^{m+1}$ onto the subspace defined by the N most important eigenvectors $\boldsymbol{\phi}_i$, i.e.

$$\mathbf{u}^{m+1} \approx \sum_{i=1}^{N} \boldsymbol{\phi}_i \; \zeta_i^{m+1} \;=\; \mathbf{B} \; \boldsymbol{\zeta}^{m+1}. \quad (8)$$

Equation (7) then writes

$$\mathbf{G}^m \; \mathbf{B} \; \boldsymbol{\zeta}^{m+1} \;=\; \mathbf{H}^m, \quad (9)$$

or equivalently

$$\mathbf{B}^T\mathbf{G}^m \; \mathbf{B} \; \boldsymbol{\zeta}^{m+1} \;=\; \mathbf{B}^T\mathbf{H}^m. \quad (10)$$

Coefficients $\boldsymbol{\zeta}^{m+1}$ defining the solution of the reduced-order model are thus obtained by solving an algebraic system of size N instead of M. When $N \ll M$, as is the case in numerous applications, the solution of Eq. (10) is thus very convenient because of its very reduced size.

Remark 2.2.1. The reduced-order model Eq. (10) is built a posteriori by means of the already-computed solution to the full-order problem. This is mandatory, in principle, in standard POD techniques. To avoid this costly solution of full-order problems, some alternatives can be found in the literature (see e.g. Bialecki et al. (2005), Burkardt et al. (2006), Gunzburger et al. (2007), Maday and Ronquist (2004), Park and Cho (1996), Ryckelynck (2003), Ryckelynck (2005), Ryckelynck (2008)). The first approach consists in solving the original full-order model over a short time interval, to extract the characteristic structure that defines the reduced model, and to hope it serves well for approximating the solution at larger time intervals. The second approach consists in solving the original, full-order, model over the entire time interval, and then use the so obtained reduced approximation to solve similar problems in which, for example, slight variations in material parameters or boundary conditions are made.

2.3 A practical example

Let us now consider the following one-dimensional heat transfer problem (see Chinesta et al. (2011)), written in non-dimensional form:

$$\frac{\partial u}{\partial t} = \lambda \frac{\partial^2 u}{\partial x^2}, \tag{11}$$

with constant thermal diffusivity $\lambda = 0.01$, $t \in (0, 30]$ and $x \in (0, 1)$. The initial condition is $u(x, t = 0) = 1$ and the boundary conditions are given by $\frac{\partial u}{\partial x}\big|_{x=0,t} = q(t)$ and $\frac{\partial u}{\partial x}\big|_{x=1,t} = 0$.

Eq. (11) is solved by means of an implicit-in-time finite element method built on a mesh of $M = 100$ nodes. Linear consistency is assumed for the finite element model. Time step is set to $\Delta t = 0.1$. The resulting algebraic system of equations reads:

$$\mathbf{K}\ \mathbf{u}^{m+1} = \mathbf{M}\ \mathbf{u}^{m} + \mathbf{q}^{m+1}. \tag{12}$$

Consider first the following heat source:

$$q(t) = \begin{cases} 1 & t \leq 10 \\ 0 & t > 10 \end{cases}. \tag{13}$$

The resulting FEM temperature profiles are shown in Fig. 1 at time instants $t_m = m$, for $m = 1, 2, ..., 30$. Red curves correspond to the heating process, occurring until $t = 10$. Afterwards, blue curves represent the solution for $t > 10$.

From these 30 discrete temperature profiles, matrices $\mathbf{Q}$ and $\mathbf{c}$ can be built in order to construict the eigenvalue problem (2). The 3 largest eigenvalues are found to be $\alpha_1 = 1790$, $\alpha_2 = 1.1$, $\alpha_3 = 0.1$, being the remaining eigenvalues $\alpha_j < \alpha_1 \cdot 10^{-8}$, $4 \leq j \leq 100$. It is expected that a reduced model made after a linear combination of the mentioned 3 eigenvectors related to the first 3 largest eigenvalues should be able to approximate the solution with sufficient accuracy. Although not strictly necessary (and even if the resulting basis will not be orthogonal), to deal with the initial condition it is very convenient to include the initial condition in the approximation basis. Figure 2 shows the resulting approximation functions after normalization, i.e. $\frac{\boldsymbol{\phi}_j}{\|\boldsymbol{\phi}_j\|}$ $(j = 1, 2, 3)$ and $\frac{\mathbf{u}^0}{\|\mathbf{u}^0\|}$. Defining the matrix $\mathbf{B}$ as

$$\mathbf{B} = \left[\frac{\mathbf{u}^0}{\|\mathbf{u}^0\|}\ \frac{\boldsymbol{\phi}_1}{\|\boldsymbol{\phi}_1\|}\ \frac{\boldsymbol{\phi}_2}{\|\boldsymbol{\phi}_2\|}\ \frac{\boldsymbol{\phi}_3}{\|\boldsymbol{\phi}_3\|} \right], \tag{14}$$

we obtain the reduced model related to Eq. (12),

$$\mathbf{B}^T\mathbf{K}\mathbf{B}\ \boldsymbol{\zeta}^{m+1} = \mathbf{B}^T\mathbf{M}\mathbf{B}\ \boldsymbol{\zeta}^{m} + \mathbf{B}^T\mathbf{q}^{m+1}, \tag{15}$$

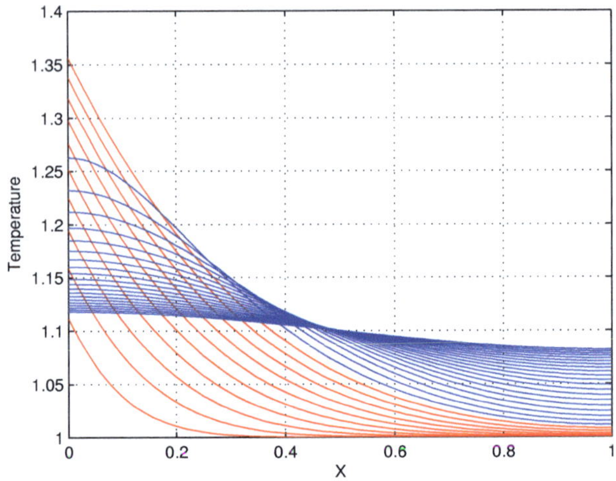

Figure 1. Temperature profiles corresponding to the problem given by Eq. (13) at time steps $t_m = m$, for $m = 1, 2, ..., 30$. Red curves correspond to the heating step, occurring until $t = 10$, while blue curves represent subsequent time step, for $t > 10$. Note the heat transfer by conduction from the warmest zones towards the cold ones.

which involves 4 degrees of freedom only. The initial condition in the reduced basis is $(\boldsymbol{\zeta}^0)^T = (1, 0, 0, 0)$.

Equation (15) and the relationship $\mathbf{u}^{m+1} = \mathbf{B}\boldsymbol{\zeta}^{m+1}$ then provide with approximate solution profiles at a very low computational cost. Results are shown in Figure 3 and are practically undistinguishable at first sight from those of the complete problem (12).

Coming back to Remark 1, assume now the usage of the reduced model (15) *as such* to solve a problem *different* from the original one. For instance, we assume now instead of (13) a substantially different heat source on the boundary:

$$q(t) = \begin{cases} \frac{t}{20} & t \leq 20 \\ \frac{t-30}{5} & t > 20 \end{cases} . \tag{16}$$

Again, the solution of the reduced model is compared to that of the complete problem in Figure 4. Even though the reduced approximation basis functions are those obtained from the thermal model related to the bound-

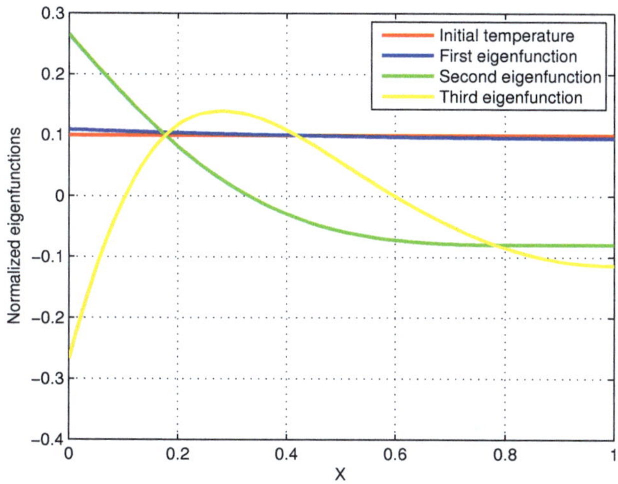

Figure 2. Reduced-order approximation basis involving the initial condition and the eigenvectors corresponding to the three largest eigenvalues.

ary condition (13), the reduced model yields a very accurate representation of the solution of this rather different problem.

2.4 Discussion

We have discussed so far the inherent advantages of model order reduction. Ideally, it would be optimal to be able to develop reduced models *a priori*, i.e. without the need for solving the complete problem. One possibility is to assess the accuracy of the reduced-order solution and, if necessary, to enrich the reduced approximation basis in order to improve accuracy (see e.g. our earlier works in the field, (Ammar et al., 2006b) and (Ryckelynck, 2005)). Proper Generalized Decomposition (PGD), which we describe in general terms in the next section, is also a very appealing alternative to these questions.

To root the origins of PGD, which will be deeply described later on, note that the previous results indicate that an accurate approximate, low-dimensional, solution to the problem can often be written as a finite sum of separated functions (here, of time and space). Indeed, when the solution evolves smoothly, as in heat transfer problems, for instance, the magnitude of the (ordered) eigenvalues α_i decreases very fast with increasing index i,

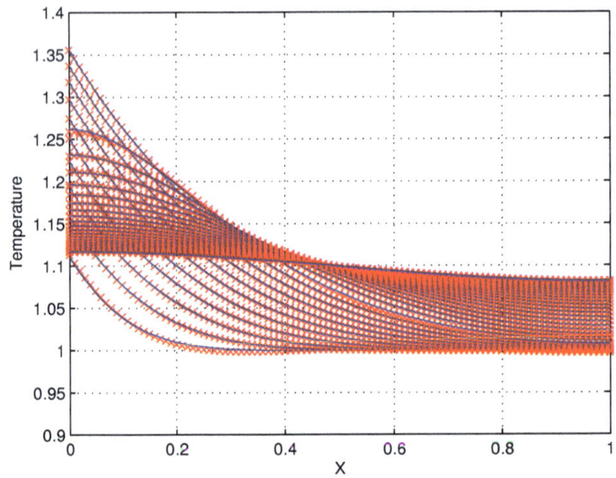

Figure 3. Global (continuous line) versus reduced-order (symbols) model solutions.

and the evolution of the field can be approximated from a very reduced number of functions in a Ritz style. Thus, if we define a threshold value ϵ (e.g. $\epsilon = 10^{-8} \cdot \alpha_1$, α_1 being the highest eigenvalue), only a small number N of modes are retained ($N \ll M$) such that $\alpha_i \geq \epsilon$, for $i \leqslant N$, and $\alpha_i < \epsilon$, for $i > N$. Equivalently:

$$u(\mathbf{x}, t) \approx \sum_{i=1}^{N} \phi_i(\mathbf{x}) \cdot T_i(t) \equiv \sum_{i=1}^{N} X_i(\mathbf{x}) \cdot T_i(t). \tag{17}$$

For the sake of clarity, the space modes $\phi_i(\mathbf{x})$ will be hereafter denoted as $X_i(\mathbf{x})$. Equation (17) rhas the form of a *separated representation*, also known as finite sum decomposition. Solutions depending on space and time can be efficiently approximated (in a great amount of cases) as a sum of a *small number* of functional products, depending on space and time, respectively. The usage of separated representations like (17) constitutes the main ingredient of PGD, the second one being the a priori nature of the method.

These functional products are determined (since we look for products of functions we face a non-linear problem) by means of iterative algorithms.The origins of this technique can be traced back to the so-called radial approximation introduced by Ladeveze (1999), Ladeveze et al. (2010), Nouy (2010)

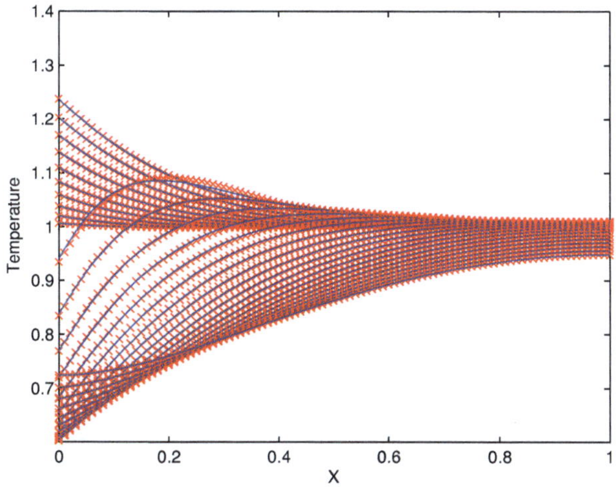

Figure 4. Global (continuous line) versus reduced-order (symbols) model solutions for the source term (16). The reduced-order approximation basis is that obtained from the solution of a different thermal problem, with the source term (13).

in the context of Computational Solid Mechanics and within the so-called LArge Time INcrement (LATIN) method.

In terms of computer cost, savings allowed by these techniques are very important. If we consider a transient problem (parabolic, hyperbolic) defined in a 3D space, the use of a standard incremental strategy with P time steps (P is often in the order of millions nowadays) requires the solution of P three-dimensional problems. On the contrary, using the a space-time separated representation (17), it is necessary to solve $N \cdot m$ three-dimensional problems to determine the spatial basis functions $X_i(\mathbf{x})$, and $N \cdot m$ more one-dimensional problems to determine the time approximation $T_i(t)$. By m we refer to the number of non-linear iterations needed for computing each term of the finite sum. Very often, it is found that $N \cdot m$ is of order 100. Computing time savings obtained by a separated representation can be of several orders of magnitude.

To this topic will be devoted the second chapter of this book. Prior to that, some aspects of POD remain to be discussed in order to fully highlight the advantages and limitations of the technique.

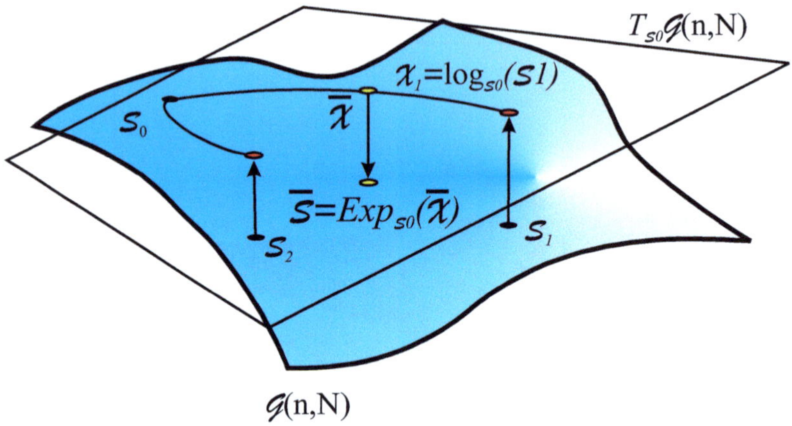

Figure 5. Schematic description of the Grassman manifold formed by the set of basis functions of all the reduced modes of a given model.

3 Interpolation of the reduced basis: a geometrical approach

One of the main limitations of POD-based reduced models is that of interpolating reduced models. Imagine that we construct a reduced approximation for a full-order problem subjected, say, to a load in a particular position. Consider now a second position of the load. Is there a relationship between these two reduced models, so that an efficient reduced-order model could be stablished for a third position of the load by interpolating between the two initial approximations?

The response is, obviously, affirmative. Amsallem and Farhat (2008) have pointed out that the set of reduced basis functions given by Eq. (2) for a given full-order model constitute the so-called Grassman manifold $\mathcal{G}(n, N)$. Therefore, in order to interpolate this set of basis functions, that obviously do not form a vector space, we must project to the tangent plane at a point of the manifold, which is a "flat" space, interpolate there, and project back to the manifold, as schematically explained in Fig. 5.

In this way, the columns of $\mathbf{B}$ constitute a basis of the subspace $\mathcal{S}_0$ of dimension n of the space $\Re^N$. At each point $\mathcal{S}$ of the manifold $\mathcal{G}(n, N)$ one can define a tangent plane of the same dimension, $\mathcal{T}_{\mathcal{S}}$, with its points defined by a matrix $\mathbf{\Gamma} \in \Re^{N\times n}$. The exponential mapping $Exp_{\mathcal{S}}$ transforms χ in an n-dimensional subspace $\mathcal{S}'$ given by a matrix $\mathbf{B}' \in \Re^{N\times n}$, such that

$$\mathbf{\Gamma} = \mathbf{U}\mathbf{\Sigma}\mathbf{V}^T \quad \text{(Singular value decomposition)}$$

$$\mathbf{B}' = \mathbf{B}\mathbf{V}\cos\boldsymbol{\Sigma} + \mathbf{U}\sin\boldsymbol{\Sigma}$$

Conversely, the logarithmic mapping $Log_{\mathcal{S}}$, defines a map between a point in the neighborhood of $\mathcal{S} \in \mathcal{G}(n, N)$ and the tangent plane at the origin. Thus, the image of $\mathcal{S}'$, in a neighborhood of $\mathcal{S}$, given by the logarithmic mapping, $\chi = Log_{\mathcal{S}}\mathcal{S}' \in \mathcal{T}_{\mathcal{S}}$ will be

$$(\mathbf{I} - \mathbf{B}\mathbf{B}^T)\mathbf{B}'(\mathbf{B}^T\mathbf{B}')^{-1} = \mathbf{U}\boldsymbol{\Sigma}\mathbf{V}^T \quad \text{(Singular value decomposition)}$$

$$\boldsymbol{\Gamma} = \mathbf{U}\tan^{-1}(\boldsymbol{\Sigma})\mathbf{V}^T$$

So, consider, for instance, $\mathbf{B}_0 \in \Re^{N\times n}$ and $\mathbf{B}_1 \in \Re^{N\times n}$, two matrices representing two subspaces, obtained for different parameters of the model (for instance, load positions, but the theory is completely general for other parameters of the model), s_0 and s_1. Let $\mathcal{S}_0$, $\mathcal{S}_1$ be the two subspaces originated by considering parameters s_0 and s_1. Let, in turn, $\mathcal{Y}(t)$ be the geodesic line that joins both subspaces (points in the Grassmann manifold), having $\mathcal{S}_0$ as origin. In that case, the initial derivative of the geodesic line, that belongs to the tangent plane at $\mathcal{S}_0$, will be

$$\dot{\mathcal{Y}}_0 = Log_{\mathcal{S}_0}\mathcal{S}_1$$

such that the matrix representing this initial derivative of the geodesic will be (see Amsallem and Farhat (2008) for a complete proof of this)

$$(\mathbf{I} - \mathbf{B}_0\mathbf{B}_0^T)\mathbf{B}_1(\mathbf{B}_0^T\mathbf{B}_1)^{-1} = \mathbf{U}\boldsymbol{\Sigma}\mathbf{V}^T$$

$$\boldsymbol{\Gamma} = \mathbf{U}\tan^{-1}(\boldsymbol{\Sigma})\mathbf{V}^T$$

Let $\tilde{\mathcal{S}}$ denote the point of the Grassmann manifold representing the reduced-order basis for the new value of the parameter. $\tilde{s}$. The initial derivative of the new geodesic line, joining $\mathcal{S}_0$ and the sought interpolated subspace $\tilde{\mathcal{S}}$, will be

$$\dot{\tilde{\mathcal{Y}}}_0 = \tilde{r}\dot{\mathcal{Y}}_0$$

with

$$\tilde{r} = \frac{\tilde{s} - s_0}{s_1 - s_0}$$

The computation of the singular value decomposition is not a very time-consuming task. For instance, on a PC which has 2GHz CPU, the computation of SVD(B) using Matlab, with **B** a random matrix of 5×5000 elements takes on average much less than 1 ms, still compatible with real-time constraints.

3.1 POD with interpolation (PODI)

A less rigorous, but much simpler, method to interpolate among previously computed reduced models was established in Ly and Tran (2005). Although in the standard PODI technique the POD procedure is applied to the complete set of snapshots (for different load positions, say) of the system to obtain an orthonormal basis $\mathbf{B} = [\phi_1 \cdots \phi_n]$, in this thesis we proceed by just applying the POD to each complete model (i.e., to each load position). Thus, we obtain an orthonormal basis for each system's parameter value. Basis are then interpolated for intermediate positions of the load. Although it is clear that the interpolation of orthogonal sets of functions does not yield, in general, to new orthogonal basis functions, the technique works well if the "distance" of reduced models in the Grassman manifold is not too large.

Results obtained with this technique showed to be much more efficient and accurate than those obtained with the more rigorous technique base upon interpolation on the Grassman manifold. This surprising result is still not well understood by the authors and is currently one of our research topics.

4 POD for non-linear models

One key aspect in the reduction of non-linear models is that incremental algorithms need for the evaluation of updated tangent stiffness matrices, and this is done in the full model. Therefore, one fundamental aspect of model reduction, the use of matrices with very limited size, is lost. This limitation is currently one of the most active research topics within the field of model order reduction. To fully comprehend this aspect, consider the simulation of the palpation of human cornea, considered in Niroomandi et al. (2008). In it, a non-linear hyperelastic model for the corneal tissue was considered.

The mesh consisted of 8514 nodes and 7182 hexahedral elements. A view of the geometry is shown in Fig. 6.

Usually, the corneal tissue is modeled as a fibrous-reinforcement hyperelastic medium. These fibre orientations are shown in Fig. 7.

Obviously, the first possibility is to test, for a single position of the load, the accuracy of the reduced model versus the complete one. As introduced before, once the complete model is solved, the most important eigenmodes are extracted from the computed displacement field, together with the initial tangent stiffness matrix (although other approaches can be envisaged, such as an intermediary tangent matrix, for instance). For the time being, let us consider that no updating of the tangent matrix is possible,so as to

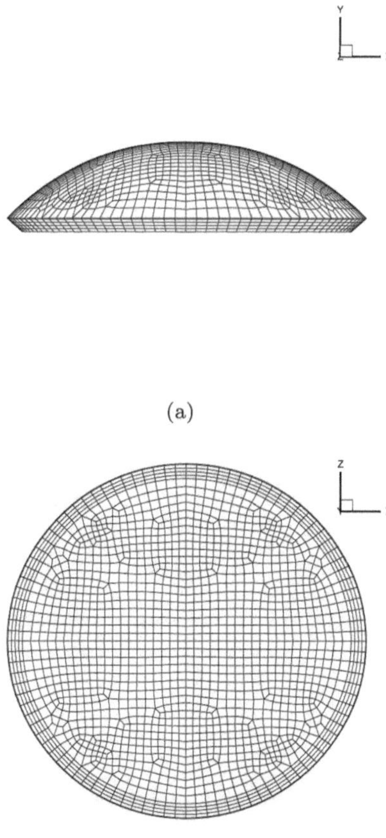

(a)

(b)

Figure 6. Geometry of the finite element model for the human cornea.

keep the computational cost as low as possible. The number of eigenmodes employed in this case was only six, leading to acceptable results. Modes are shown in Fig. 8. Their respective eigenvalues are, from the biggest to the smallest one, $9.02 \cdot 10^4$, 690, 27, 2.63, 0.221 and 0.0028. The relative influence of the eignemodes on the solution decays very fast, as can be noticed from the eigenvalues. Compare also the 8514 nodes with three degrees of freedom each, thus making 25542 degrees of freedom, of the full model, with the six (three-dimensional) degrees of freedom of the reduced model. The computational saving is enormous.

To fully compare the results, different positions of the load were con-

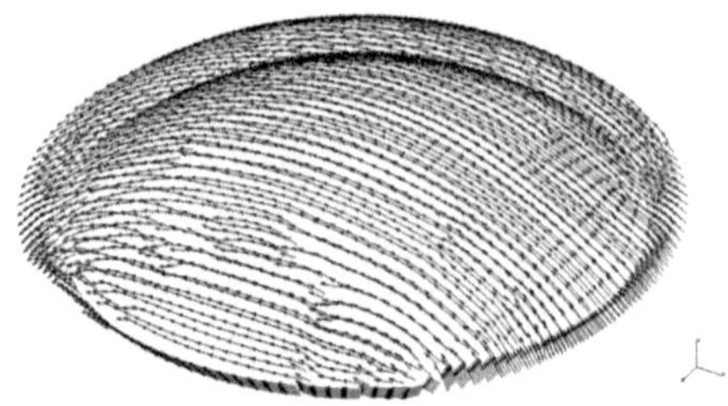

(a)

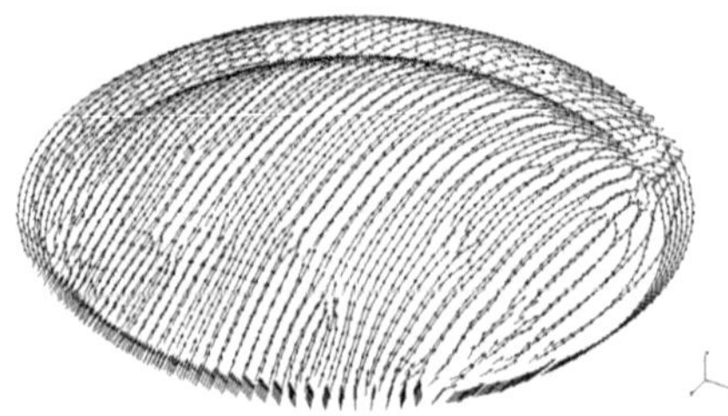

(b)

Figure 7. Fiber distribution in the finite element model of the cornea. Two families of collagen fibers, roughly perpendicular to each other, are considered.

sidered. It was found that different positions do not very much influence the level of accuracy, roughly the same for any loading position. For an arbitrary location of the load, the obtained vertical displacement is shown in Fig. 9.

In Fig. 10 load-displacement curve has been depicted. In Fig. 11, however, the load was applied at a point located slightly towards the outer boundary of the model. In this case, as can be seen from Fig. 11, the displacement obtained at the point of application of the load is nearly exact, although the shape of the deformed cornea is somewhat different. This is not the case for Fig. 9, where errors of about 20% are noticed. The L_2 error norm ranged from very low values (0.08) in the early steps of the simulation, to higher values (around 0.34) for the last step. In our experience, this is a typical upper bound of the obtained error, even if very large deformations are imposed to the simulated organ, as is the case.

Notice that, due to the shape of shallow dome of the cornea, severe buckling phenomena are expected. This highly non-linear behavior can not be captured by a reduced model in which no tangent stiffness matrix

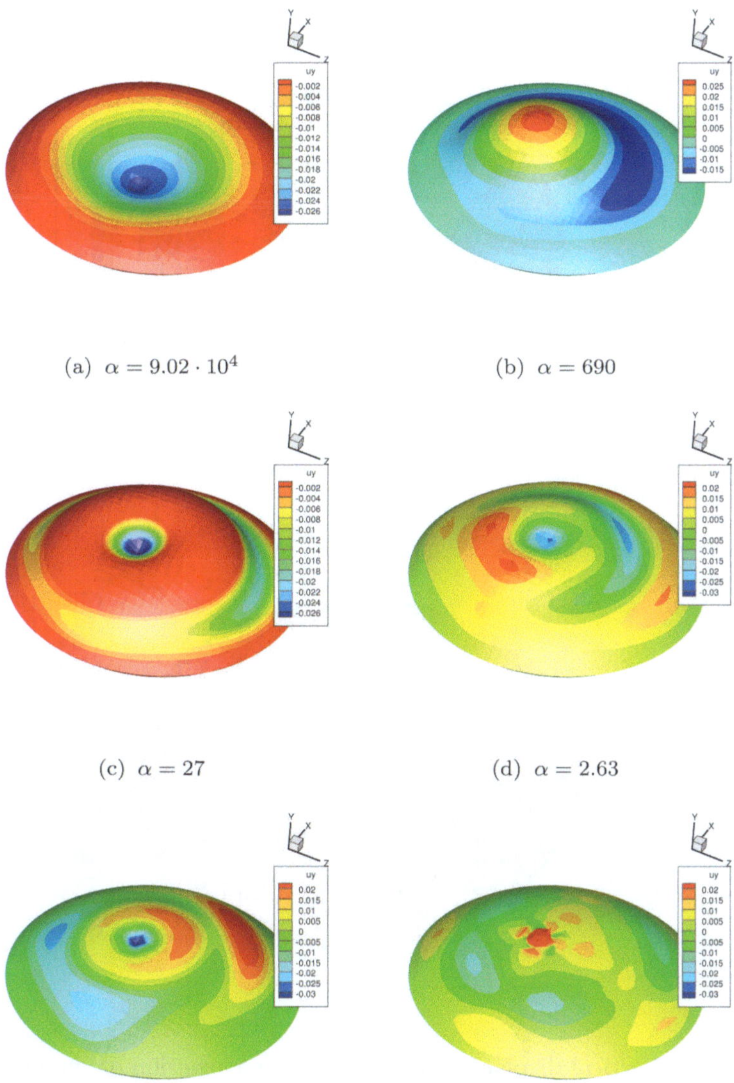

(a) $\alpha = 9.02 \cdot 10^4$

(b) $\alpha = 690$

(c) $\alpha = 27$

(d) $\alpha = 2.63$

(e) $\alpha = 0.221$

(f) $\alpha = 0.0028$

Figure 8. The eigenmodes of the problem employed as global basis for the reduced model simulation.

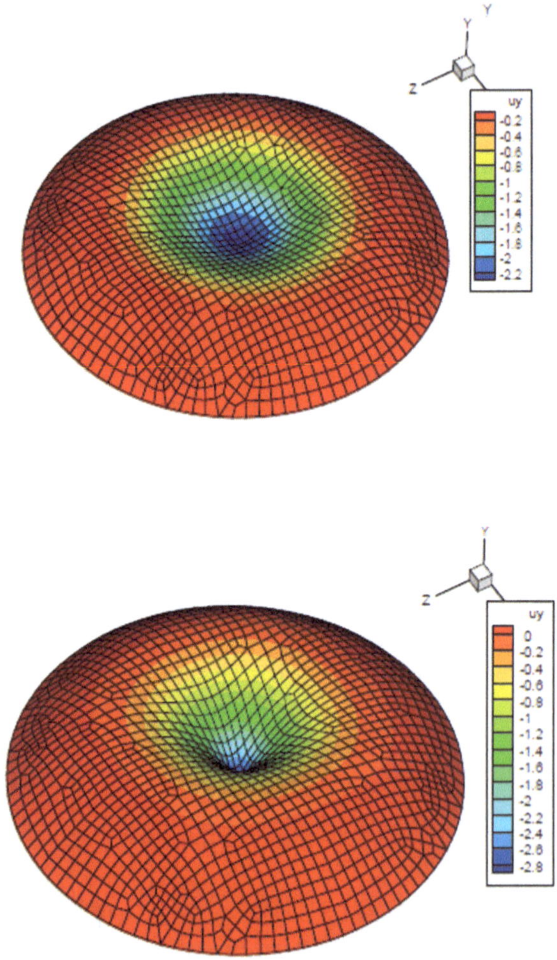

Figure 9. Vertical displacement field for a first position of the load. Complete model (up) vs. reduced model (down).

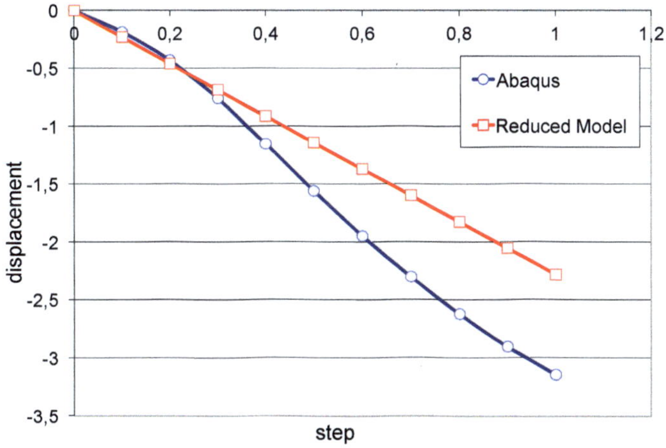

Figure 10. load-displacement curve at the vertex of the cornea.

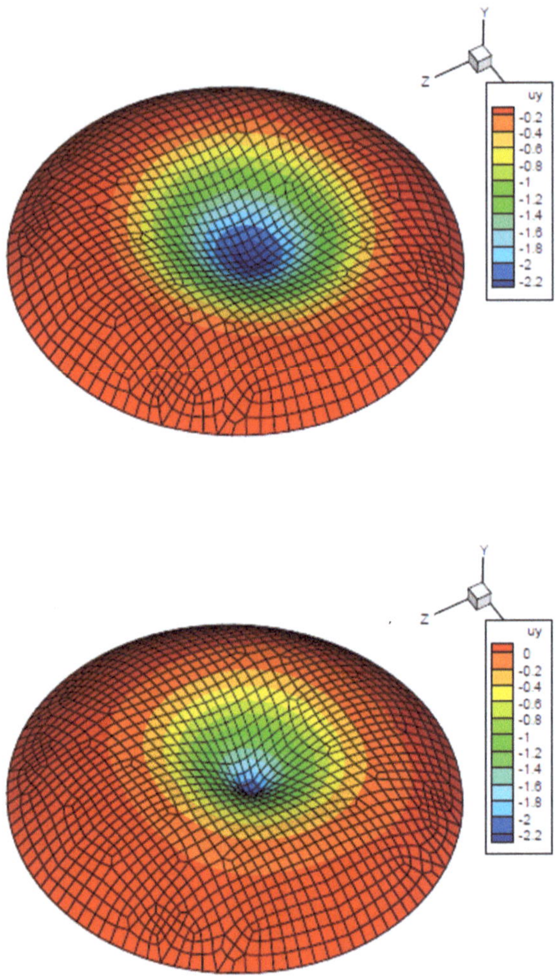

Figure 11. Vertical displacement field for a second position of the load. Complete model (up) vs. reduced model (down).

is allowed, of course. However, all the simulations presented here ran on a PC equipped with two processors (only one was employed, no parallel computing was used) AMD Quad Opteron running at 2.2 GHz and with 16 Gb RAM, under Scientific Linux. The prototype code was implemented under MATLAB.

The simulations ran at 472-483Hz, compliant with the highest requirements for haptic feedback, for instance.

4.1 Force located at an arbitrary point

Of course, one of the key aspects of the construction of a suitable reduced model based on POD is that of obtaining a good basis. Snapshots should be taken from representative states (here, load states) of the system. It is therefore of utmost importance to know the quality of the response of the system to a force located in a position whose response has not been calculated. In this work, we employ the PODI approach mentioned before (Ly and Tran, 2005). Therefore, the worst case scenario is to apply a load between two loads for which the complete, full-order, solution is known, and to compare it with a full order solution.

Under very large strains, the results of the full order simulation are shown in Fig. 12, while results for the PODI model are, as can be noticed, in good agreement, see Fig. 12(b).

The error in the prediction of the vertical displacement under the load is 27.18% at the end of the simulation (maximum of the strain). Error is computed in the $||\cdot||_2$ norm, defined as:

$$||e||_2 = \frac{1}{n}\sqrt{\sum_{I=1}^{n} e_I^2} \tag{18}$$

where e_I represents the nodal error and n the number of nodes in the model. Maximum error took a value of 29.5%, still within the limits for the best techniques available today for linear elastic materials (Lim and De, 2007).

4.2 Other approaches to the problem

Despite this classical approach to non-linear problems (simply to ignore the non-linearities), many solutions have been proposed recently in the literature of reduced order models. One of the possibilities is to employ POD in conduction with Asymptotic Numerical Methods (ANM), a technique due to M. Potier-Ferry and coworkers (Cochelin et al., 1994a; Abichou et al., 2002; Cao and Potier-Ferry, 1999; Cochelin et al., 1994b).

Let us introduce in a simple manner the ANM. Consider a very simple non-linear problem:

$$\frac{\partial u}{\partial t} - k\Delta u = u^2 + f(\mathbf{x}, t)$$

The non linear term can be affected by a "loading" parameter λ

$$\frac{\partial u}{\partial t} - k\Delta u = \lambda \cdot u^2 + f(\mathbf{x}, t) \tag{19}$$

We denote by u_0 the solution related to $\lambda = \lambda_0 = 0$ that can be computed easily because it corresponds to the solution of the linear problem. The sought solution is the one related to $\lambda = 1$. The ANM method is based upon defining an asymptotic expansion of the unknown field u as well as of the loading parameter λ by considering the powers of a:

$$\begin{cases} u = u_0 + a \cdot u_1 + a^2 \cdot u_2 \ + \cdots \\ \lambda = \lambda_0 + a \cdot \lambda_1 + a^2 \cdot \lambda_2 \ + \cdots \end{cases} \tag{20}$$

This non linear term can be written as:

$$u^2 = (u^2)_0 + a \cdot (u^2)_1 + a^2 \cdot (u^2)_2 \ + \cdots + a^p \cdot (u^2)_p + \cdots$$

where $(u^2)_p$ reads:

$$(u^2)_p = \sum_{i=0}^{i=p} u_i \cdot u_{p-i} = 2 \cdot u_0 \cdot u_p + \sum_{i=1}^{i=p-1} u_i \cdot u_{p-i} \tag{21}$$

Introducing (20) and (21) into (19) and identifying the different powers of a, it results:

- Power $p = 0$. Solution u_0 is assumed known because it corresponds to the solution of the linear problem resulting from $\lambda = \lambda_0 = 0$.
- Power $p = 1$. We should solve:

$$\frac{\partial u_1}{\partial t} - k\Delta u_1 = \lambda_0 \cdot (u^2)_1 + \lambda_1 \cdot (u^2)_0$$

  and since $\lambda_0 = 0$, the evolution equation for u_1 reduces to:

$$\frac{\partial u_1}{\partial t} - k\Delta u_1 = \lambda_1 \cdot (u^2)_0 \tag{22}$$

The previous equation contains two unknowns, u_1 and λ_1, so to obtain a solution we need for an extra equation. In Cochelin et al. (1994a) the following is considered:

$$(u - u_0, u_1) + (\lambda - \lambda_0, \lambda_1) = a \tag{23}$$

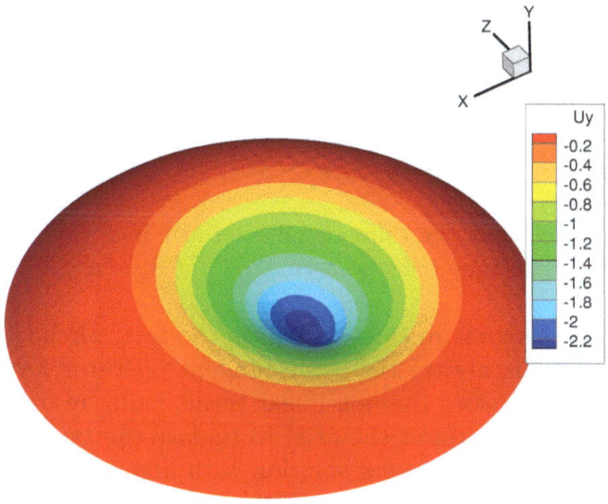

(a)

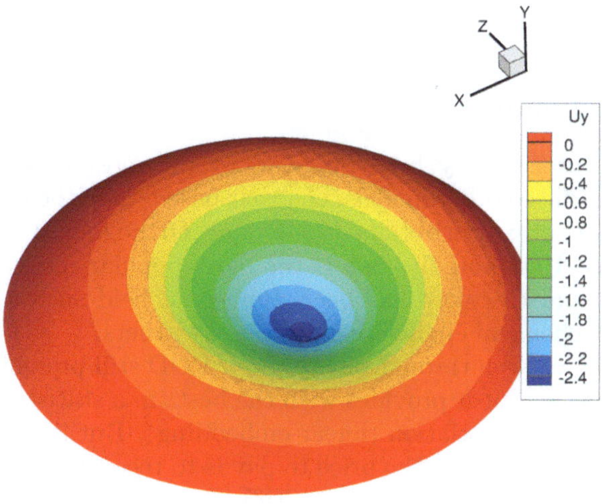

(b)

Figure 12. Vertical displacement (mm) for (a) the reduced model and (b) the complete model of a load in a position not previously computed.

where $(\cdot, \cdot)$ denotes the scalar product.

By algebraic considerations only, we arrive at an expression for the solution at power p, with $p > 1$:

$$\frac{\partial u_p}{\partial t} - k\Delta u_p = \lambda_p \cdot (u^2)_0 + \sum_{i=1}^{i=p-1} \lambda_i \cdot (u^2)_{p-i} \tag{24}$$

where we notice that the differential operator is always the same and that in the right hand member the term $\sum_{i=1}^{i=p-1} \lambda_i \cdot (u^2)_{p-i}$ has been already computed.

What is important in this method is the fact, already mentioned, that Eq. (24) contains always the same tangent operator for the term at any order of the expansion, the non-linear terms being re-directed to the force vector. Therefore, without the need to update the tangent stiffness matrix we arrive at a truly non-linear solution with high accuracy, see Niroomandi et al. (2010b).

Other approaches to this same problem are based upon the same philosophy, i.e., to interpolate the non-linear term of the equation. To do so, an appropriate set of points should be chosen so as to provide the best possible interpolated results. The so-called Empirical Interpolation Method (Chaturantabut and Sorensen, 2010) or the Best Points Interpolation Method (Nguyen et al., 2008) pertain to this same approach, although they differ in some aspects.

5 Conclusions

In this chapter a very brief overview of the most classical technique to construct reduced order models, i.e., that based upon the projection on a low dimensional subspaces obtained by Proper Orthogonal Decomposition, has been made. Undoubtedly, this is the most popular model order reduction technique, and therefore has been applied in field such as solid and fluid mechanics, chemical engineering, and many others.

However, as seen through the chapter, POD still presents two main difficulties. One os that it provides with the best (in a statistical sense) basis to perform a linear simulation (linear embedding). But if we deal with a non-linear problems, usually the price to pay is to reconstruct the reduced order matrix by updating the full order stiffness matrix of the original problem. Therefore, many of the computational savings are lost.

The second crucial aspects is that of constructing a good sample of snapshots, i.e., points at which the full order model is evaluated so as to attract the eigenmodes to be employed as basis functions in a Ritz-like framework. No definitive answer has been given to those tow big problems of POD.

Dealing with the construction of *a priori* reduced order models (i.e., those without the need for the solution of full order problems), we must cite the works by Ryckelynck et al. (2006) and, notably, the appearance of Proper Generalized Decompositions (PGD), developed independently by Ladeveze (1999) and Ammar et al. (2006a, 2007). PGD methods, that will be introduced in subsequent chapters of this book, allow for an a priori development of reduced order models without the need for the solution of complete ones, and therefore have constituted a little revolution in the field.

With respect to the other big problem, that of non-linear problems, several possibilities have been briefly overviewed in this chapter. This same problematic is common to PGD techniques. In general, either an approach based upon asymptotic expansions (Cochelin et al., 1994a) or one based on the interpolation of the non-linear terms are preferred (Chaturantabut and Sorensen, 2010; Nguyen et al., 2008).

In any case, model order reduction techniques have become a powerful technique in situations in which many-query frameworks are present (optimization, parametric problems, ...) or where real-time responses are mandatory (real-time simulation, (Niroomandi et al., 2010a), for instance).

Bibliography

H. Abichou, H. Zahrouni, and M. Potier-Ferry. Asymptotic numerical method for problems coupling several nonlinearities . *Computer Methods in Applied Mechanics and Engineering*, 191(51-52):5795–5810, 2002.

A. Ammar, B. Mokdad, F. Chinesta, and R. Keunings. A new family of solvers for some classes of multidimensional partial differential equations encountered in kinetic theory modeling of complex fluids. *J. Non-Newtonian Fluid Mech.*, 139:153–176, 2006a.

A. Ammar, D. Ryckelynck, F. Chinesta, and R. Keunings. On the reduction of kinetic theory models related to finitely extensible dumbbells. *Journal of Non-Newtonian Fluid Mechanics*, 134(1-3):136 – 147, 2006b. ISSN 0377-0257. doi: http://dx.doi.org/10.1016/j.jnnfm.2006.01.007.

A. Ammar, B. Mokdad, F. Chinesta, , and R. Keunings. A new family of solvers for some classes of multidimensional partial differential equations encountered in kinetic theory modeling of complex fluids. part ii: transient simulation using space-time separated representations. *J. Non-Newtonian Fluid Mech.*, 144:98–121, 2007.

D. Amsallem and C. Farhat. An interpolation method for adapting reduced-order models and application to aeroelasticity. *AIAA Journal*, 46:1803–1813, 2008.

R. A. Bialecki, A. J. Kassab, and A. Fic. Proper orthogonal decomposition and modal analysis for acceleration of transient fem thermal analysis. *International Journal for Numerical Methods in Engineering*, 62 (6):774–797, 2005. ISSN 1097-0207. doi: 10.1002/nme.1205. URL http://dx.doi.org/10.1002/nme.1205.

John Burkardt, Max Gunzburger, and Hyung-Chun Lee. {POD} and cvt-based reduced-order modeling of navier?stokes flows. *Computer Methods in Applied Mechanics and Engineering*, 196(1?3):337 – 355, 2006. ISSN 0045-7825. doi: http://dx.doi.org/10.1016/j.cma.2006.04.004.

H.-L. Cao and M. Potier-Ferry. An improved iterative method for large strain viscoplastic problems. *International Journal for Numerical Methods in Engineering*, 44:155–176, 1999.

Saifon Chaturantabut and Danny C. Sorensen. Nonlinear model reduction via discrete empirical interpolation. *SIAM Journal on Scientific Computing*, 32(5):2737–2764, 2010. ISSN 1064-8275. doi: 10.1137/090766498.

F. Chinesta, A. Ammar, A. Leygue, and R. Keunings. An overview of the proper generalized decomposition with applications in computational rheology. *Journal of Non-Newtonian Fluid Mechanics*, 166(11):578 – 592, 2011. ISSN 0377-0257. doi: http://dx.doi.org/10.1016/j.jnnfm.2010.12.012. ¡ce:title¿XVIth International Workshop on Numerical Methods for Non-Newtonian Flows¡/ce:title¿.

B. Cochelin, N. Damil, and M. Potier-Ferry. Asymptotic-numerical methods and Padé approximants for non-linear elastic structures. *International Journal for Numerical Methods in Engineering*, 37:1187–1213, 1994a.

B. Cochelin, N. Damil, and M. Potier-Ferry. The asymptotic numerical method: an efficient perturbation technique for nonlinear structural mechanics. *Revue Europeenne des Elements Finis*, 3:281–297, 1994b.

Max D. Gunzburger, Janet S. Peterson, and John N. Shadid. Reduced-order modeling of time-dependent {PDEs} with multiple parameters in the boundary data. *Computer Methods in Applied Mechanics and Engineering*, 196(4?6):1030 – 1047, 2007. ISSN 0045-7825. doi: http://dx.doi.org/10.1016/j.cma.2006.08.004.

P. Ladeveze. *Nonlinear Computational Structural Mechanics*. Springer, N.Y., 1999.

P. Ladeveze, J.-C. Passieux, and D. Neron. The {LATIN} multiscale computational method and the proper generalized decomposition. *Computer Methods in Applied Mechanics and Engineering*, 199(21?22):1287 – 1296, 2010. ISSN 0045-7825. doi: http://dx.doi.org/10.1016/j.cma.2009.06.023. ¡ce:title¿Multiscale Models and Mathematical Aspects in Solid and Fluid Mechanics¡/ce:title¿.

Yi-Je Lim and Suvranu De. Real time simulation of nonlinear tissue response in virtual surgery using the point collocation-based method of finite spheres. *Computer Methods in Applied Mechanics and Engineering*, 196:3011–3024, 2007.

Hung V. Ly and Hien T. Tran. Modeling and control of physical processes using proper orthogonal decomposition. *Mathematical and Computer Modelling*, 33:223–236, 2005.

Y. Maday and E. Ronquist. The reduced basis element method: Application to a thermal fin problem. *SIAM Journal on Scientific Computing*, 26(1):240–258, 2004. doi: 10.1137/S1064827502419932. URL `http://epubs.siam.org/doi/abs/10.1137/S1064827502419932`.

N. C. Nguyen, A. T. Patera, and J. Peraire. A 'best points' interpolation method for efficient approximation of parametrized functions. *International Journal for Numerical Methods in Engineering*, 73(4):521–543, JAN 22 2008. ISSN 0029-5981. doi: 10.1002/nme.2086.

S. Niroomandi, I. Alfaro, E. Cueto, and F. Chinesta. Real-time deformable models of non-linear tissues by model reduction techniques. *Computer Methods and Programs in Biomedicine*, 91(3):223–231, 2008.

S. Niroomandi, I. Alfaro, E. Cueto, and F. Chinesta. Accounting for large deformations in real-time simulations of soft tissues based on reduced-order models. *Computer Methods and Programs in Biomedicine*, In Press, Corrected Proof:–, 2010a. ISSN 0169-2607. doi: DOI: 10.1016/j.cmpb.2010.06.012.

Siamak Niroomandi, Iciar Alfaro, Elias Cueto, and Francisco Chinesta. Model order reduction for hyperelastic materials. *International Journal for Numerical Methods in Engineering*, 81(9):1180–1206, 2010b. ISSN 1097-0207. doi: 10.1002/nme.2733. URL `http://dx.doi.org/10.1002/nme.2733`.

Anthony Nouy. A priori model reduction through proper generalized decomposition for solving time-dependent partial differential equations. *Computer Methods in Applied Mechanics and Engineering*, 199(23-24):1603 – 1626, 2010. ISSN 0045-7825. doi: http://dx.doi.org/10.1016/j.cma.2010.01.009.

H. M. Park and D. H. Cho. The use of the Karhunen-Loève decomposition for the modeling of distributed parameter systems. *Chemical Engineering Science*, 51(1):81–98, 1996.

D. Ryckelynck. A priori model reduction method for the optimization of complex problems. In *Workshop on Optimal Design of Materials and Structures, Ecole Polytechnique, Palaiseau, Paris (France)*, 2003.

D. Ryckelynck. A priori hyperreduction method: an adaptive approach. *Journal of Computational Physics*, 202(1):346–366, 2005.

D. Ryckelynck. Hyper-reduction of mechanical models involving internal variables. *International Journal for Numerical Methods in Engineering*, 77(1):75–89, 2008.

D. Ryckelynck, F. Chinesta, E. Cueto, and A. Ammar. On the a priori Model Reduction: Overview and recent developments. *Archives of Computational Methods in Engineering*, 12(1):91–128, 2006.

PGD for solving multidimensional and parametric models

Francisco Chinesta[*], Elias Cueto[†] and Antonio Huerta[‡]

[*] ESI Group Chair on Advanced Computational Manufacturing Processes
GEM, UMR CNRS - Centrale Nantes
1 rue de la Noe, BP 92101, F-44321 Nantes cedex 3, France

[†] I3A, Universidad de Zaragoza
Maria de Luna s/n, 50018 Zaragoza, Spain

[‡] Laboratori de Calcul Numeric
Universidad Politecnica de Cataluña
Jordi Girona 1, Campus Nord, C2, E-08034 Barcelona, Spain

Abstract In this chapter we are addressing a new paradigm in the field of simulation-based engineering sciences (SBES) to face the challenges posed by current ICT technologies. Despite the impressive progress attained by simulation capabilities and techniques, some challenging problems remain today intractable. These problems, that are common to many branches of science and engineering, are of different nature. Among them, we can cite those related to high-dimensional problems, which do not admit mesh-based approaches due to the exponential increase of degrees of freedom. We developed in recent years a novel technique, called Proper Generalized Decomposition (PGD). It is based on the assumption of a separated form of the unknown field and it has demonstrated its capabilities in dealing with high-dimensional problems overcoming the strong limitations of classical approaches. But the main opportunity given by this technique is that it allows for a completely new approach for classic problems, not necessarily high dimensional. Many challenging problems can be efficiently cast into a multidimensional framework and this opens new possibilities to solve old and new problems with strategies not envisioned until now. For instance, parameters in a model can be set as additional extra-coordinates of the model. In a PGD framework, the resulting model is solved once for life, in order to obtain a general solution that includes all the solutions for every possible value of the parameters, that is, a sort of *computational vademecum*. Under this rationale, optimization of complex problems, uncertainty quantification, simulation-based control and real-time simulation are now at hand, even in highly complex scenarios, by combining an off-line stage in which

F. Chinesta, P. Ladevèze (Eds.), *Separated Representations and PGD-Based Model Reduction Fundamentals and Applications*, CISM International Centre for Mechanical Sciences
DOI 10.1007/ 978-3-7091-1794-1_2 © CISM Udine 2014

the general PGD solution, the vademecum, is computed, and an on-line phase in which, even on deployed, handheld, platforms such as smartphones or tablets, real-time response is obtained as a result of our queries.

1 Introduction

1.1 Motivation

Six unique initiatives have been recently selected (and funded with 100 millions of euros per year) by the European Research Council based solely on their potential for realizing scientific breakthroughs and influencing Europe's social and industrial challenges, including health. Their aim will then be to deliver major breakthroughs in information and communication technologies (ICT), with the potential to provide solutions to some of society's biggest challenges. Despite being different there is a common ingredient to all of them, which is to emphasize the necessity of making use of advanced simulation-driven sciences and engineering, as will be highlighted below. The six contenders, from which the two flagship initiatives were selected, are: (http://www.epractice.eu/en/news/5304734)

1. Guardian Angels for a Smarter Life project aimed at developing tiny devices without batteries that act like thinking and autonomous personal assistants, providing information and communication technologies to assist people in all sorts of complex situations delivering features and characteristics that go well beyond human capabilities. http://www.ga-project.eu/
2. The Human Brain Project whose goal is to understand the way the human brain works. The long-term goal of the Human Brain Project is to build the informatics, modeling, and supercomputing technologies that are needed to simulate and understand the human brain. http://www.ga-project.eu/
3. IT Future of Medicine proposes a data-driven, individualized medicine of the future, based on the molecular / physiological / anatomical data from individual patients. The project outcomes will enable data-driven real-time calculation of health, disease, therapy and its effects for individual patients. http://www.itfom.eu/
4. Robot Companions for Citizens is a project devoted to developing soft-skinned and intelligent robots with highly developed perceptive, cognitive and emotional skills. Robot Companions for Citizens will be based on the novel solid articulated structures with flexible properties displaying soft behavior, haptic devices and simulation based real time

control in deployed systems. These companions will also have new levels of perceptual, cognitive and emotive capabilities and be aware of their physical and social surroundings and respond accordingly. http://www.itfom.eu/

5. FuturICT Knowledge Accelerator and Crisis-Relief System. What if global scale computing facilities were available that could analyze most of the data available in the world? What insights could scientists gain about the way society functions? What new laws of nature would be revealed? Could society discover a more sustainable way of living? ICT (Information and Communication Technology) can analyze vast amounts of data and complex situations so as to better predict natural disasters, or manage and respond to man-made disasters that cross national borders or continents.
http://www.futurict.eu
6. Graphene Science and technology for ICT and beyond. Graphene is a new substance developed by atomic and molecular scale manipulation that could replace silicon as the wonder material of the 21st century. This aims to explore revolutionary potentials, in terms of both conventional as well as radically new fields of Information and Communication Technologies applications.
http://www.graphene-flagship.eu/

It is now well known (http://www.itfom.eu/) that the human brain consumes 4 watts for performing some tasks that today's computers will require the power of several nuclear plants. It is then clear that our computers and algorithms for addressing the models encountered in science and engineering are definitively suboptimal. The above six flagship projects share some key aspects related to efficient computational sciences. It is expected that these projects will reach a certain number of breakthroughs, but all of them will face important limitations of today's computer capabilities and, notably, simulation techniques.

All these society needs require fast and accurate solutions, in general data-driven, of very complex models, involving an unimaginable amount of information, in most cases in real time and on deployed platforms. Up to now, the solution of complex models, preferably fast and accurate, is addressed by using high performance computing and hyper powerful computing platforms. Obviously the consecution of the above "dreams" will require as much as computational power (supercomputing) as possible, and consequently, advances in hardware and software for high-performance computing will be necessary. But at the same time, there is a need for a new generation simulation techniques, beyond high-performance computing or nowadays approaches (most of them proposed 40 years ago), to simply improve efficiency or to allow getting results when other alternatives fail in the above challenging scenarios.

All the above challenging problems are data-driven. The importance of Dynamic Data-Driven Application Systems –DDDAS – in the forthcoming decades has been already noticed by the NSF Blue Ribbon Panel on Simulation Based Engineering Sciences report, that in 2006 included DDDAS as one of the five core issues or challenges in the field for the next decade (together with multi-scale simulation, model validation and verification, handling large data and visualization). This panel concluded that "Dynamic data-driven application systems will rewrite the book on the validation and verification of computer predictions" and that "research is needed to effectively use and integrate data-intensive computing systems, ubiquitous sensors and high-resolution detectors, imaging devices, and other data-gathering storage and distribution devices, and to develop methodologies and theoretical frameworks for their integration into simulation systems" (see NSF Final Report, DDDAS Workshop 2006, Arlington, VA, U.S.A. (2006); Darema (1994); Oden et al. (2006)). Moreover, the NSF believes that "... The DDDAS community needs to reach a critical mass both in terms of numbers of investigators, and in terms of the depth, breadth and maturity of constituent technologies ...'

1.2 Nowadays computational issues

Today many problems in science and engineering remain intractable, in spite of the impressive progresses attained in modeling, numerical analysis, discretization techniques and computer science during the last decade, because their numerical complexity, or the restrictions imposed by different requirements (real-time on deployed platforms, for instance) make them unaffordable for today's technologies.

We can enumerate different challenging scenarios for efficient numerical simulations:

- The first one concerns models that are defined in high dimensional spaces, usually encountered in quantum chemistry describing the structure and mechanics of materials (Cances et al. (2003); Ammar et al. (2008)), the kinetic theory description of complex materials (Bird et al. (1987); Lamari et al. (2012)), social dynamics and economic systems, vehicular traffic flow phenomena, complex biological systems involving mutation and immune competition, crowds and swarms encountered in congested and panic flows, among many other unimaginable possibilities (see Bellomo (2008) and the references therein); the chemical modeling in too dilute systems where the concept of concentration cannot be used, that results in the so-called chemical master equation governing for example cell signaling and other phenomena in molecular biology (Ammar, Cueto and Chinesta (2012)).

Models defined in high dimensional spaces suffer the so-called curse of dimensionality. If one proceeds to the solution of a model defined in a space of dimension D by using a standard mesh based discretization technique, where M nodes are used for discretizing each space coordinate, the resulting number of nodes reaches the astronomical value of M^D. With $M \approx 10^3$ (a very coarse description in practice) and $D \approx 30$ (a very simple model) the numerical complexity results 10^{90}. It is important to recall that 10^{80} is the presumed number of elementary particles in the universe!.

Traditionally, high dimensional models were addressed by using stochastic simulations. However these techniques have their own challenges: variance reduction is always an issue and the construction of distribution functions in high dimensional spaces remains in most cases unaffordable. It is also quite difficult within the stochastic framework to implement parametric or sensitivity analysis that go beyond the brute force approach of computing a large number of expensive, individual simulations.

- Online control can be carried out following different approaches. The most common one consists in considering systems as a black box whose behavior is modeled by a transfer function relating certain inputs to certain outputs. This modeling that may seem poor has as main advantage the possibility of proceeding rapidly due to its simplicity. This compromise between accuracy and rapidity was often used in the past and this pragmatic approach has allowed us to control processes and to optimize them, once the transfer function modeling the system is established.

  The establishment of such goal-oriented transfer function is the trickiest point. For this purpose, it is possible to proceed from a sometimes overly simplified physical model or directly from experiments (allowing us to extract a phenomenological goal-oriented transfer function) or from a well-balanced mixture of both approaches. In all cases, the resulting modeling can only be applied within the framework that served to derive it. However, on one hand, the fine description of systems requires a sufficiently detailed description of them and, in that case, traditional goal-oriented simplified modeling becomes inapplicable. On the other hand, actual physical models result, in general, in complex mathematical objects, non-linear and strongly coupled partial differential equations. Such mathematical objects are representing physical reality up to a certain degree of accuracy. However, the available numerical tools capable of solving these complex models require the use of powerful computers that can require hours, days and weeks to solve them. Known as numerical simulation, its output solution is very rich but it seems inapplicable for control purposes that require fast responses, often in real-time.

Until now, numerical simulation has been used offline but in some cases it allows us to define simplified models (with their inherent limitations and drawbacks) running in real-time that could be used online but such simplified modeling has the previously quoted drawbacks.

- Many problems in parametric modeling, inverse identification, and process or shape optimization, usually require, when approached with standard techniques, the direct computation of a very large number of solutions of the concerned model for particular values of the problem parameters. When the number of parameters increases such a procedure becomes inapplicable.
- Traditionally, Simulation-based Engineering Sciences - SBES - relied on the use of static data inputs to perform the simulations. These data could be parameters of the model(s) or boundary conditions. The word static is intended to mean here that these data could not be modified during the simulation. A new paradigm in the field of Applied Sciences and Engineering has emerged in the last decade. Dynamic Data-Driven Application Systems (DDDAS) constitute nowadays one of the most challenging applications of simulation-based Engineering Sciences. By DDDAS we mean a set of techniques that allow the linkage of simulation tools with measurement devices for real-time control of simulations. DDDAS entails the ability to dynamically incorporate additional data into an executing application, and in reverse, the ability of an application to dynamically steer the measurement process.

  In this context, real time simulators are needed in many applications. One of the most challenging situations is that of haptic devices, where forces must be translated to the peripheral device at a rate of 500 Hz. Control, malfunctioning identification and reconfiguration of malfunctioning systems also need to run in real time. All these problems can be seen as typical examples of DDDAS.
- Augmented reality is another area in which efficient (fast and accurate) simulation is urgently needed. The idea is supplying in real time appropriate information to the reality perceived by the user. Augmented reality could be an excellent tool in many branches of science and engineering. In this context, light computing platforms are appealing alternatives to heavy computing platforms that in general are expensive and whose use requires technical knowledge.

- Inevitable uncertainty. In science and engineering, in its widest sense, it now seems obvious that there are many causes of variability. The introduction of such variability, randomness and uncertainty is a priority for the next decade. Although it was a priority in the preceding decade, the practical progress attained seems fairly weak.

While the previous list is by no means exhaustive, it includes a set of problems with no apparent relationship between them that can however be treated in a unified manner as will be shown in what follows. Their common ingredient is our lack of capabilities (or knowledge) to solve them efficiently in a direct, traditional way.

1.3 Fast calculations from a historical perspective

The human being throughout the history developed several facilities for giving fast responses to a variety of questions. Thus, abaci were used 2700 years B.C. in Mesopotamia. This abacus was a sort of counting frame primarily used for performing arithmetic calculations. We associate this abacus to a bamboo frame with beads sliding on wires, however, originally they were beans or stones moved in grooves in sand or on tablets of wood, stone, or metal. The abacus was in use centuries before the adoption of the written modern numeral system and is still widely used by readers. There are many variants, the Mesopotamian abacus, the Egyptian, Persian, Greek, Roman, Chinese, Indian, Japanese, Korean, native American, Russian, etc.

However, the initial arithmetic needs were rapidly complemented with more complex representations. We are considering some few variants:

- Charts appeared for graphical representation of data with multiple meanings. However, there are common features that provide the chart with its ability to extract meaning from data. In general a chart is graphical, containing very little text, since humans infer meaning from pictures quicker than from text. A particular variant of charts in the Nomogram.
- Nomography, is the graphical representation of mathematical relationships or laws. It is an area of practical and theoretical mathematics invented in 1880 by Philbert Maurice d'Ocagne and used extensively for many years to provide engineers with fast graphical calculations of complicated formulas to a practical precision. Thus, a nomogram can be considered as a graphical calculating device. There are thousands of examples on the use of nomograms in all the fields of sciences and engineering.

The former facilities allowed for fast calculations and data manipulations. Nomograms can be easily constructed when the mathematical relationships that they express are purely algebraic, eventually non-linear. In those cases it was easy to represent some outputs as a function of some inputs. The calculation of these data representations was performed off-line and then used on-line in many branches of engineering sciences for design and optimization.

However, the former procedures fail when addressing more complex scenarios. Thus, sometimes engineers manipulate not properly understood physics and in that case the construction of nomograms based on a too coarse modelling could be dangerous. In that cases one could proceed by making several experiments from which defining a sort of experiment-based nomogram. In other cases the mathematical object to be manipulated consists of a system of complex coupled non-linear partial differential equations, whose solution for each possible combination of the values of the parameters that it involves is simply unimaginable for the nowadays computational availabilities. In these cases experiments or expensive computational solutions are performed for some possible states of the system, from which a simplified model linking the inputs to the outputs of interest is elaborated. These simplified models have different names: surrogate models, metamodels, response surface methodologies, ... Other associated tricky questions are the one that concerns the best sampling strategy (Latin hypercube, ...) and also the one concerning the appropriate interpolation techniques for estimating the response at an unmeasured position from observed values at surrounding locations. Many possibilities exist, being Kriging one of the most widely used for interpolating data.

All these techniques allow defining a sort of numerical or graphical handbook. One of the earliest and most widely known within engineering practice is that of Bernoulli (1836). However, we must accept a certain inevitable inaccuracy when estimating solutions from the available data. It is the price to pay if neither experimental measurements nor numerical solutions of the fine but expensive model are achievable for each possible scenario.

Recently model order reduction opened new possibilities. First, proper orthogonal decompositions (POD) allows extracting the most significant characteristic of the solution, that can be then applied for solving models slightly different to the ones that served to defined the reduced approximation bases. There is an extensive literature. The interested readers can reefer to Park and Cho (1996); Maday and Ronquist (2004); Bialecki (2005); Burkardt (2006); Gunzburger (2007); Ryckelynck (2003); Ammar, Ryckelynck et al. (2006); Niroomandi et al. (2008), (2010) and (2012); Ammar, Pruliere, Chinesta et al. (2009); Schmidt et al. (2011) and the numerous references therein. The extraction of the reduced basis is the tricky point when using POD-based model order reduction, as well its adaptivity when addressing scenarios far from the ones considered when constructing the reduced basis (Ryckelynck (2005); Ryckelynck et al. (2005) and (2006)). Another issue lies in the error control, and its connection with verification and validation.

The calculation of the reduced basis is not unique. There are many alternatives. Some ones introduce some improvements on the POD methodology, as is the case of the Goal Oriented Model Constrained Optimization approach (see Bui-Thanh et al. (2007) and the references therein) or the modal identification method (see Girault et al. (2010) and the references therein). The Branch Eigenmodes Reduction Method combined with the amalgam method is another appealing constructor of reduced bases (Videcoq et al. (2010))

Another family of model reduction techniques lies in the used of reduced basis constructed by combining a greedy algorithm and a priori error indicator. It needs for some amount off-line work but then the reduced basis can be used on-line for solving different models with a perfect control of the solution accuracy because the availability of error bounds. When the error is inadmissible, the reduced basis can be enriched by invoking again the same greedy algorithm. The interested reader can refer to Maday et al. (2002); Veroy and Patera (2005); Rozza et al. (2008); Eftang et al. (2011) and the references therein. The main drawback of such an approach is the amount of data that must be computed, stored and then manipulated.

1.4 Model reduction: information versus relevant information

Model Order Reduction techniques have been extensively used, as just indicated, for solving transient models defined in space domains involving a large number of degrees of freedom, small time steps and a large time interval.

Imagine a model whose solution $u(\mathbf{x},t)$ is defined in $\mathbf{x} \in \Omega$ with $t \in \mathcal{I} = (0,T]$. By using an appropriate discretization technique the solution is calculated at M points $\mathbf{x}_i$, $i = 1,\cdots,M$ and P time instants t_m, $m = 1,\cdots,P$, being the number of nodes M and time instants P large enough for ensuring convergence and stability of the discrete solution.

Standard discretization techniques (finite elements, finite volumes, finite differences, meshless techniques, ...) require in general the solution of a linear system of size M at each time instant. Thus the computational complexity scales with $P \cdot M$. When M becomes large because the solution localization or the domain complexity, or/and P because stability constraints or because the size of the time interval with respect to the characteristic time of the physics involved, the resulting complexity $P \cdot M$ could become unaffordable.

Model reduction techniques based on the use of Proper Orthogonal Decomposition allowed circumventing the just referred computational difficulty when the problem solution lives in a subspace of dimension much smaller than M.

Extracting relevant information: the Proper Orthogonal Decomposition

We assume that the field of interest $u(\mathbf{x},t)$ is known at the nodes $\mathbf{x}_i$ of a spatial mesh for discrete times $t_m = m \cdot \Delta t$, with $i \in [1, \cdots, M]$ and $m \in [0, \cdots P]$. We use the notation $u(\mathbf{x}_i, t_m) \equiv u^m(\mathbf{x}_i) \equiv u_i^m$ and define $\mathbf{u}^m$ as the vector of nodal values u_i^m at time t_m. The main objective of the POD is to obtain the most typical or characteristic structure $\phi(\mathbf{x})$ among these $u^m(\mathbf{x})$, $\forall m$ (Ryckelynck et al. (2006)). For this purpose, we maximize the scalar quantity

$$\alpha = \frac{\sum_{m=1}^{P} \left[\sum_{i=1}^{M} \phi(\mathbf{x}_i) u^m(\mathbf{x}_i) \right]^2}{\sum_{i=1}^{M} (\phi(\mathbf{x}_i))^2}, \tag{1}$$

which amounts to solve the following eigenvalue problem:

$$\mathbf{C}\boldsymbol{\phi} = \alpha\boldsymbol{\phi}. \tag{2}$$

Here, the vector $\boldsymbol{\phi}$ has i-component $\phi(\mathbf{x}_i)$, and $\mathbf{C}$ is the two-point correlation matrix

$$\mathbf{C}_{ij} = \sum_{m=1}^{P} u^m(\mathbf{x}_i) u^m(\mathbf{x}_j) \tag{3}$$

whose compact form writes

$$\mathbf{C} = \sum_{m=1}^{P} \mathbf{u}^m \cdot (\mathbf{u}^m)^T, \tag{4}$$

which is symmetric and positive definite. With the matrix $\mathbf{Q}$ defined as

$$\mathbf{Q} = \begin{pmatrix} u_1^1 & u_1^2 & \cdots & u_1^P \\ u_2^1 & u_2^2 & \cdots & u_2^P \\ \vdots & \vdots & \ddots & \vdots \\ u_M^1 & u_M^2 & \cdots & u_M^P \end{pmatrix}, \tag{5}$$

we have

$$\mathbf{C} = \mathbf{Q} \cdot \mathbf{Q}^T. \tag{6}$$

In order to obtain a reduced model, we first solve the eigenvalue problem Eq. (2) and select the N eigenvectors $\boldsymbol{\phi}_i$ associated with the eigenvalues belonging to the interval defined by the highest eigenvalue α_1 and α_1 divided by a large enough number (e.g. 10^8). In practice, for many models, N is found to be much lower than M. These N eigenfunctions $\boldsymbol{\phi}_i$ are then used to approximate the solution $u^m(\mathbf{x})$, $\forall m$. To this end, let us define the matrix $\mathbf{B} = [\boldsymbol{\phi}_1 \cdots \boldsymbol{\phi}_N]$, i.e.

$$\mathbf{B} = \begin{pmatrix} \phi_1(\mathbf{x}_1) & \phi_2(\mathbf{x}_1) & \cdots & \phi_N(\mathbf{x}_1) \\ \phi_1(\mathbf{x}_2) & \phi_2(\mathbf{x}_2) & \cdots & \phi_N(\mathbf{x}_2) \\ \vdots & \vdots & \ddots & \vdots \\ \phi_1(\mathbf{x}_M) & \phi_2(\mathbf{x}_M) & \cdots & \phi_N(\mathbf{x}_M) \end{pmatrix}. \tag{7}$$

Now, let us assume for illustrative purposes that an explicit time-stepping scheme is used to compute the discrete solution $\mathbf{u}^{m+1}$ at time t^{m+1}. One must thus solve a linear algebraic system like

$$\mathbf{G}^m \ \mathbf{u}^{m+1} = \mathbf{H}^m. \tag{8}$$

A reduced-order model is then obtained by approximating $\mathbf{u}^{m+1}$ in the subspace defined by the N eigenvectors $\boldsymbol{\phi}_i$, i.e.

$$\mathbf{u}^{m+1} \approx \sum_{i=1}^{N} \boldsymbol{\phi}_i \ T_i^{m+1} = \mathbf{B} \ \mathbf{T}^{m+1}. \tag{9}$$

Equation (8) then reads

$$\mathbf{G}^m \ \mathbf{B} \ \mathbf{T}^{m+1} = \mathbf{H}^m, \tag{10}$$

or equivalently

$$\mathbf{B}^T \mathbf{G}^m \ \mathbf{B} \ \mathbf{T}^{m+1} = \mathbf{B}^T \mathbf{H}^m. \tag{11}$$

The coefficients T_i^{m+1} defining the solution of the reduced-order model are thus obtained by solving an algebraic system of size N instead of M. When $N \ll M$, as is the case in numerous applications, the solution of Eq. (11) is thus preferred because of its much reduced size.

Towards an "a priori" space-time separated representation

The above example illustrates the significant value of model reduction. Of course, one would ideally want to be able to build a reduced-order approximation *a priori*, i.e. without relying on the knowledge of the (approximate) solution of the complete problem. One would then want to be able to assess the accuracy of the reduced-order solution and, if necessary, to enrich the reduced approximation basis in order to improve accuracy (see e.g. our earlier studies: Ammar, Ryckelynck et al. (2006) and Ryckelynck et al. (2006)). The Proper Generalized Decomposition (PGD), which we describe in general terms in the next section, is an efficient answer to these questions.

The above POD results also tell us that an accurate approximate solution can often be written as a separated representation involving but few terms. Indeed, when the field evolves smoothly, the magnitude of the (ordered) eigenvalues α_i decreases very fast with increasing index i, and the evolution of the field can be approximated from a reduced number of modes. Thus, if we define a cutoff value ϵ (e.g. $\epsilon = 10^{-8} \cdot \alpha_1$, α_1 being the highest eigenvalue), only a small number N of modes are retained ($N \ll M$) such that $\alpha_i \geq \epsilon$, for $i \leqslant N$, and $\alpha_i < \epsilon$, for $i > N$. Thus, one can write:

$$u(\mathbf{x}, t) \approx \sum_{i=1}^{N} \phi_i(\mathbf{x}) \cdot T_i(t) = \sum_{i=1}^{N} X_i(\mathbf{x}) \cdot T_i(t). \tag{12}$$

For the sake of clarity, the space modes $\phi_i(\mathbf{x})$ will be denoted in the sequel as $X_i(\mathbf{x})$. Equation (12) represents a natural *separated representation*, also known as finite sum decomposition. The solution that depends on space and time can be approximated as a sum of a *small number* of functional products, with one of the functions depending on the space coordinates and the other one on time. Use of separated representations like (12) is at the heart of the PGD.

Separated representations were introduced in the 80s by Pierre Ladeveze that proposed a space-time separated representation of transient solutions involved in strongly non-linear models, defining a non-incremental integration procedure. The interested reader can refer to the numerous Ladeveze's works: Ladeveze (1989) and (1999); Ladeveze et al. (2002), (2003), (2007), (2010) and (2011); Neron and Ladeveze (2010) and Passieux et al. (2010).

In terms of performance, the verdict is simply impressive. Consider a typical transient problem defined in 3D physical space. Use of a standard incremental strategy with P time steps (P could be of order of millions in industrial applications) requires the solution of P three-dimensional problems. By contrast, using the space-time separated representation (12), we must solve $N \cdot m$ three-dimensional problems for computing the space functions $X_i(\mathbf{x})$, and $N \cdot m$ one-dimensional problems for computing the time functions $T_i(t)$. Here, m is the number of non-linear iterations needed for computing each term of the finite sum. For many problems of practical interest, we find that $N \cdot m$ is of order 100. The computing time savings afforded by the separated representation can thus reach many orders of magnitude.

Later, separated representations were employed in the context of stochastic modelling (Nouy (2010)) as well as for solving multidimensional models suffering the so-called curse of dimensionality, some of them never solved before (Ammar, Mokdad et al. (2006) and (2007)). The techniques making use of separated representations computed on the fly were called Proper Generalized Decompositions – PGD –.

PGD constitutes an efficient multidimensional solver that allows introducing model parameters (boundary conditions, initial conditions, geometrical parameters, material and process parameters ...) as extra-coordinates. Then by solving only once and off-line the resulting multidimensional model we have access to the parametric solution that can be viewed as a sort of handbook or vademecum than can be then used on-line.

In what follows, we are describing within the PGD approach the way of introducing extra-coordinates of different nature. Later, we will prove the potentiality of such an approach for the efficient solution of a variety of problems.

1.5 The Proper Generalized Decomposition at a glance

Consider a problem defined in a space of dimension D for the unknown field $u(x_1, \cdots, x_D)$. Here, the coordinates x_i denote any usual coordinate (scalar or vectorial) related to physical space, time, or conformation space, for example, but they could also include problem parameters such as boundary conditions or material parameters as described later. We seek a solution for $(x_1, \cdots, x_D) \in \Omega_1 \times \cdots \times \Omega_D$.

The PGD yields an approximate solution in the separated form:

$$u(x_1, \cdots, x_D) \approx \sum_{i=1}^{N} F_i^1(x_1) \cdot \ldots \cdot F_i^d(x_D). \tag{13}$$

The PGD approximation is thus a sum of N functional products involving each a number D of functions $F_i^j(x_j)$ that are unknown *a priori*. It is constructed by successive enrichment, whereby each functional product is determined in sequence. At a particular enrichment step $n+1$, the functions $F_i^j(x_j)$ are known for $i \leq n$ from the previous steps, and one must compute the new product involving the D unknown functions $F_{n+1}^j(x_j)$. This is achieved by invoking the weak form of the problem under consideration. The resulting discrete system is non-linear, which implies that iterations are needed at each enrichment step. A low-dimensional problem can thus be defined in Ω_j for each of the D functions $F_{n+1}^j(x_j)$. The interested reader can refer to the primer (Chinesta et al. (2014)) for a valuable introduction on the PGD technology.

If M nodes are used to discretize each coordinate, the total number of PGD unknowns is $N \cdot M \cdot D$ instead of the M^D degrees of freedom involved in standard mesh-based discretizations.

2 Separated representations in action

In this section we are describing the way in which separated representation can be used for circumventing or alleviating the challenges enumerated in the first section.

2.1 Multidimensional models

We consider the model

$$\mathcal{L}(u) = f(x_1, \cdots, x_D) \tag{14}$$

defined in $\Omega_1 \times \cdots \times \Omega_D$ with $u(x_1, \cdots, x_D)$. In the previous equation $\mathcal{L}$ represents the differential operator involving the D coordinates that for the sake of simplicity, and without loss of generality, is assumed linear and accepting the following decomposition

$$\mathcal{L}(u) = \sum_{i=1}^{i=D} \mathcal{G}_i(x_1, \cdots, x_D) \cdot \mathcal{L}_{x_i}(u) \tag{15}$$

where $\mathcal{L}_{x_i}$ denotes a linear differential operator with respect to the x_i coordinate.

The solution procedure consists of assuming the separated representation

$$u(x_1, \cdots, x_D) \approx \sum_{i=1}^{N} F_i^1(x_1) \cdot ... \cdot F_i^D(x_D). \tag{16}$$

and insert it in the weak form associated with Eq. (14)

$$\int_{\Omega_1 \times \cdots \times \Omega_D} u^* \cdot (\mathcal{L}(u) - f) \ dx_1 \ \cdots \ dx_D = 0 \tag{17}$$

As just described the solution is constructed by successive enrichments, whereby each functional product is determined in sequence. At a particular enrichment step n, the functions $F_i^j(x_j)$ are known for $i < n$ from the previous steps, and one must compute the new product involving the D unknown functions $F_n^j(x_j)$.

As the problem of calculating the D functions $F_n^j(x_j)$ at enrichment step n is nonlinear, the use of an appropriate linearization scheme is mandatory.

The simplest strategy widely considered in our works consists of using an alternated direction fixed point algorithm in which $F_n^j(x_j)$ is computed assuming all the other functions $F_n^k(x_k)$, $\forall k \neq j$ known. In this case the test function u^* writes

$$u^*(x_1, \cdots, x_D) = (F_n^j)^* \cdot \prod_{k=1, k\neq j}^{D} F_n^k(x_k) \tag{18}$$

By introducing Eqs. (16) and (18) into (17) and integrating in $\Omega_1 \times \cdots \times \Omega_{j-1} \times \Omega_{j+1} \times \cdots \times \Omega_D$ it results

$$\int_{\Omega_j} (F_n^j)^* \cdot \left(\alpha^j \cdot \mathcal{L}_{x_j}(F_n^j(x_j)) + \beta^j \cdot F_n^j + g_n^j(x_j)\right) \, dx_j = 0 \tag{19}$$

where α^j and β^j are two constants and $g_n^j(x_j)$ is the function that results from the integration of $f(x_1, \cdots, x_D)$ and $\mathcal{L}(\sum_{i=1}^{n-1} F_i^1(x_1) \cdot \ldots \cdot F_i^D(x_D))$.

2.2 Efficient transient non-incremental solutions

As just discussed standard incremental transient solutions require in general the solution of a 3D problem at each time step. When the number of time steps and/or the size of the space problems to be solved at each time step increase, standard incremental solutions become rapidly inefficient. POD-based model order reduction can alleviate such drawback by considering a reduced space approximation basis that allows approximating the model solution at each time step. Thus the solution procedure remains incremental but now a small linear system of equation must be solved at each time step.

The main difficulty when considering such an approach lies in the fact that despite the fact of using a reduced basis, in general, one must evaluate the whole discrete matrix $\mathbf{G}^m$ at each time step t_m (8) and then by pre and post-multiplying it by $\mathbf{B}^T$ and $\mathbf{B}$ respectively one obtain the reduced linear system (11) to be solved at reach time step t_m. While standard integration procedures spend most of the computing time in the solution of large linear systems of size M, one at each time step, POD-based reduced models are constrained by the necessity of assembling the whole system prior to solve the resulting reduced linear system.

Many possibilities exist to alleviate this task. Ryckelynck proposed the so-called hyper-reduction framework (Ryckelynck (2003)) that consists in constructing few rows of matrix $\mathbf{G}^m$. In fact as the reduced model only involves N unknowns, N equations should be enough. Even if in that case $\mathbf{G}^m$ is singular, it is not after pre and post-multiplying it by $\mathbf{B}^T$ and $\mathbf{B}$

respectively because the functions involved in the reduced basis, i.e. the rows of matrix $\mathbf{B}$, are defined in the whole space domain Ω.

The trickiest point within the hyper-reduction framework lies in the choice of these N equations. Different variants have been proposed and in general they employ slightly more equations that strictly needed in order to enhance accuracy. Other techniques (Gappy-POD, GNAT, Missing Point Estimation, ...) consider alternative strategies for the construction of the reduced discrete matrix. These techniques are not addressed in the present work but there exist a vast literature on all them.

Obviously when addressing linear models in fixed domains with constant model parameters, matrix $\mathbf{G}$ remains unchanged during the time evolution. In this case matrix $\mathbf{G}$ could be calculated and factorized only once. Thus, in this particular scenario the interest of using model order reduction is not plenty justified.

The nonlinear case is the main domain of interest for considering model reduction because except when using very particular nonlinear solvers, like the asymptotic numerical method (Cochelin et al. (1994)), one should construct and factorize the discrete matrix $\mathbf{G}^m$ at least one time at each time step t_m. In this context interpolation techniques applied on the nonlinear term allow impressive computing time savings (Barrault et al. (2004); Chaturantabut and Sorensen (2010)).

Other alternative route consists of making use of a non-incremental integration scheme based on the space-time separated representation. For illustrating it, we consider the simplest equation that using the notation introduced in the previous section reads:

$$\mathcal{L}(u) = f(\mathbf{x}, t) \tag{20}$$

defined in $\Omega \times \mathcal{I}$ with $u(\mathbf{x}, t)$. We consider, again without loss of generality, the simplest case

$$\mathcal{L}(u) = \mathcal{L}_t(u) + \mathcal{L}_x(u) \tag{21}$$

where $\mathcal{L}_t = \frac{\partial}{\partial t}$ and $\mathcal{L}_x$ is the Laplace operator, i.e. $\mathcal{L}_x = \nabla^2$.

The solution procedure consists of assuming the space-time separated representation

$$u(\mathbf{x}, t) \approx \sum_{i=1}^{N} X_i(\mathbf{x}) \cdot T_i(t) \tag{22}$$

and insert it in the weak form associated with Eq. (20)

$$\int_{\Omega \times \mathcal{I}} u^* \cdot (\mathcal{L}(u) - f) \ d\mathbf{x} \ dt = 0. \tag{23}$$

At a particular enrichment step n, functions $X_i(\mathbf{x})$ and $T_i(t)$, $i < n$, are known from the previous steps, and one must compute the new term involving functions $X_n(\mathbf{x})$ and $T_n(t)$.

As the problem of calculating functions X_n and T_n at enrichment step n is nonlinear, the use of an appropriate linearization scheme is mandatory. The simplest consists of using an alternated direction fixed point algorithm in which X_n is computed assuming T_n known, and then updating T_n from the just calculated X_n. The iteration procedure continues until reaching convergence (the fixed point).

When looking for X_n the test function u^* writes

$$u^*(\mathbf{x}, t) = X_n^* \cdot T_n. \tag{24}$$

By introducing Eqs. (24) and (22) into (23) and integrating in $\mathcal{I}$ it results

$$\int_\Omega X_n^* \cdot (\alpha^x \cdot \mathcal{L}_x(X_n(\mathbf{x})) + \beta^x \cdot X_n(\mathbf{x}) + g_n^x(\mathbf{x})) \; d\mathbf{x} = 0 \tag{25}$$

where α^x and β^x are two constants and $g_n^x(\mathbf{x})$ is the function that results from the integration in $\mathcal{I}$ of $f(\mathbf{x}, t)$ and $\mathcal{L}(\sum_{i=1}^{n-1} X_i(\mathbf{x}) \cdot T_i(t))$.

The weak form (25) can be solved by using any appropriate discretization technique for elliptic boundary value problems.

Now, when looking for T_n the test function u^* writes

$$u^*(\mathbf{x}, t) = X_n \cdot T_n^*. \tag{26}$$

By introducing Eqs. (26) and (22) into (23) and integrating in Ω it results

$$\int_\mathcal{I} T_n^* \cdot \left(\alpha^t \cdot \mathcal{L}_t(T_n(t)) + \beta^t \cdot T_n(t) + g_n^t(t)\right) \; dt = 0 \tag{27}$$

where α^t and β^t are two constants and $g_n^t(t)$ is the function that results from the integration in Ω of $f(\mathbf{x}, t)$ and $\mathcal{L}(\sum_{i=1}^{n-1} X_i(\mathbf{x}) \cdot T_i(t))$.

This equation can be solved by using any appropriate discretization technique for initial value problems.

It can be noticed that when using the non-incremental space-time separated representation instead of solving a 3D problem at each time step, we must solved of the order of N 3D problems related to the calculation of functions $X_i(\mathbf{x})$ and N one-dimensional problems for calculating functions $T_i(t)$. Even when using a extremely small time step, the computing time related to the solution of the 1D problems associated with functions $T_i(t)$ is negligible with respected to the computing time required for solving the 3D problems associated with $X_i(\mathbf{x})$. Thus finally the complexity does not depend on the time step when using a space-time separated representation and then the computing time savings can be impressive, of several orders of magnitude in many cases.

2.3 Space separation in degenerated domains

In what follows we consider a problem defined in a plate domain in which we are interested in calculating its 3D solution. In many cases when models are defined in plate or shell like domains different simplifying hypotheses can be introduced in order to reduce its 3D complexity to 2D. It is under this rationale that plate and shell elastic theories where developed in structural mechanics and lubrication models where proposed for addressing viscous flows. However in many applications the complex physics involved do not allow for such dimensional reduction and the only valuable route consist of solving the fully 3D model. However, in such degenerated domains 3D meshes involve too many elements with the associated impact in the solution procedure efficiency.

We define the 3D plate domain Ω from $\Omega = \Xi \times \Gamma$, where Γ represents the plate thickness. Thus $(x, y, z) \in \Omega$ can be expressed from $(\mathbf{x}, z)$ with $\mathbf{x} \in \Xi \subset \mathcal{R}^2$ and $z \in \Gamma \subset \mathcal{R}$.

We consider again for the sake of simplicity the 3D heat equation defined in $\Omega = \Xi \times \Gamma$

$$\mathcal{L}(u) = f(\mathbf{x}, z), \tag{28}$$

with $u(\mathbf{x}, z)$. We consider, again without loss of generality, the simplest case

$$\mathcal{L}(u) = \mathcal{L}_x(u) + \mathcal{L}_z(u), \tag{29}$$

where $\mathcal{L}_x = \frac{\partial^2}{\partial x^2} + \frac{\partial^2}{\partial y^2}$ and $\mathcal{L}_z = \frac{\partial^2}{\partial z^2}$.

The solution procedure consists in assuming an in-plane-out-of-plane separated representation

$$u(\mathbf{x}, z) \approx \sum_{i=1}^{N} X_i(\mathbf{x}) \cdot Z_i(z), \tag{30}$$

and insert it in the weak form associated with Eq. (28)

$$\int_{\Xi \times \Gamma} u^* \cdot (\mathcal{L}(u) - f) \; d\mathbf{x} \; dz = 0. \tag{31}$$

At a particular enrichment step n, functions $X_i(\mathbf{x})$ and $Z_i(z)$, $i < n$, are known from the previous steps, and one must compute the new term involving functions $X_n(\mathbf{x})$ and $Z_n(z)$.

As the problem of calculating functions X_n and Z_n at enrichment step n is nonlinear, the use of an appropriate linearization scheme is mandatory. Again, the simplest consists of using an alternated direction fixed point algorithm in which X_n is computed assuming Z_n known, and then updating Z_n from the just calculated X_n. The iteration procedure continues until reaching convergence (the fixed point).

When looking for X_n the test function u^* writes

$$u^*(\mathbf{x}, z) = X_n^* \cdot Z_n. \tag{32}$$

By introducing Eqs. (32) and (30) into (31) and integrating in Γ it results

$$\int_{\Xi} X_n^* \cdot (\alpha^x \cdot \mathcal{L}_x(X_n(\mathbf{x})) + \beta^x \cdot X_n(\mathbf{x}) + g_n^x(\mathbf{x})) \; d\mathbf{x} = 0 \tag{33}$$

where α^x and β^x are two constants and $g_n^x(\mathbf{x})$ is the function that results from the integration in Γ of $f(\mathbf{x}, t)$ and $\mathcal{L}(\sum_{i=1}^{n-1} X_i(\mathbf{x}) \cdot Z_i(z))$.

The weak form (33) can be solved by using any appropriate discretization technique for elliptic boundary value problems.

Now, when looking for Z_n the test function u^* writes

$$u^*(\mathbf{x}, t) = X_n \cdot Z_n^*. \tag{34}$$

By introducing Eqs. (34) and (30) into (31) and integrating in Ξ it results

$$\int_{\Gamma} Z_n^* \cdot (\alpha^z \cdot \mathcal{L}_z(Z_n(z)) + \beta^z \cdot Z_n(z) + g_n^z(z)) \; dz = 0, \tag{35}$$

where α^z and β^z are two constants and $g_n^z(z)$ is the function that results from the integration in Ξ of $f(\mathbf{x}, z)$ and $\mathcal{L}(\sum_{i=1}^{n-1} X_i(\mathbf{x}) \cdot Z_i(z))$.

This equation can be solved by using any appropriate discretization of 1D boundary value problems.

It can be noticed that when using the in-plane-out-of-plane separated representation instead of solving a 3D problem, we must solved of the order of N 2D problems related to the calculation of functions $X_i(\mathbf{x})$ and N one-dimensional problems for calculating functions depending of the thickness coordinate z, $Z_i(z)$. Even when using a extremely fine discretization of the domain thickness, the computing time related to the solution of the 1D problems associated with functions $Z_i(z)$ is negligible with respected to the computing time required for solving the 2D problems associated with the calculation of functions $X_i(\mathbf{x})$. Thus finally the complexity does not depend on thickness representation when using in-plane-out-of-plane separated representations and then the computing time savings can be again impressive.

2.4 Parametric solutions and computational vademecums

As emphasized in the first section of the present chapter, optimization, inverse identification and uncertainty quantification require the solution of many direct problems when considering standard procedures. In the same way real time simulation and the associated DDDAS are only possible if each solution can be performed fast enough.

There are many techniques that have been proposed in order to increase the solution efficiency. High performance computing and the recent GPUs based computations are contributing to enlarge the simulation capabilities. However and despite the impressive recent progresses many problems of industrial interest, in particular those related to decision making in complex design frameworks, remain unaffordable.

Within the separated representation framework the situation could be radically different. Due to the fact that separated representations can break the curse of dimensionality, one could envisage considering the parameters of interest (the ones involved in optimization - including geometrical parameters involved in shape optimization - or inverse identification, the ones having a stochastic distribution, or the ones concerned by real time simulation - boundary conditions, ...) as extra-coordinates. Then by solving only once and off-line the resulting multidimensional model we have access to the parametric solution that can be viewed as a sort of handbook, virtual chart or vademecum than can be then used on-line in real time and even in deployed platforms like tablets or smartphones. These vademecums can be viewed as a sort of modern monograms related to complex multiparametric partial differential equations.

In this section, we consider again the heat transfer equation but in the present case parametrized by the thermal diffusivity k

$$\mathcal{L}(u) = f(\mathbf{x}, t), \tag{36}$$

defined in $\Omega \times \mathcal{I} \times \mathcal{K}$ with $k \in \mathcal{K} \subset \mathcal{R}^+$ and $u(\mathbf{x}, t, k)$. We consider, again without loss of generality, the simplest case

$$\mathcal{L}(u) = \mathcal{L}_t(u) + k \ \mathcal{L}_x(u), \tag{37}$$

where $\mathcal{L}_t = \frac{\partial}{\partial t}$ and $\mathcal{L}_x$ is the Laplace operator, i.e. $\mathcal{L}_x = \nabla^2$.

The solution procedure consists in assuming the space-time-parameter separated representation

$$u(\mathbf{x}, t, k) \approx \sum_{i=1}^{N} X_i(\mathbf{x}) \cdot T_i(t) \cdot K_i(k), \tag{38}$$

and insert it in the weak form associated with Eq. (36)

$$\int_{\Omega \times \mathcal{I} \times \mathcal{K}} u^* \cdot (\mathcal{L}(u) - f) \; d\mathbf{x} \; dt \; dk = 0. \tag{39}$$

At a particular enrichment step n, functions $X_i(\mathbf{x})$, $T_i(t)$ and $K_i(k)$, $i < n$, are known from the previous steps, and one must compute the new term involving functions $X_n(\mathbf{x})$, $T_n(t)$ and $K_n(k)$.

As the problem of calculating functions X_n, T_n and K_n at enrichment step n is nonlinear, the use of an appropriate linearization scheme is mandatory. The simplest consists of using an alternated direction fixed point algorithm in which X_n is computed assuming T_n and K_n known, then updating T_n from the just calculated X_n and the old K_n and finally updating K_n from the just calculated X_n and T_n. The iteration procedure continues until reaching convergence, that is the fixed point.

When looking for X_n the test function u^* writes

$$u^*(\mathbf{x}, t, k) = X_n^* \cdot T_n \cdot K_n. \tag{40}$$

By introducing Eqs. (40) and (38) into (39) and integrating in $\mathcal{I} \times \mathcal{K}$ it results

$$\int_{\Omega} X_n^* \cdot (\alpha^x \cdot \mathcal{L}_x(X_n(\mathbf{x})) + \beta^x \cdot X_n(\mathbf{x}) + g_n^x(\mathbf{x})) \; d\mathbf{x} = 0, \tag{41}$$

where α^x and β^x are two constants and $g_n^x(\mathbf{x})$ is the function that results from the integration in $\mathcal{I} \times \mathcal{K}$ of $f(\mathbf{x}, t)$ and $\mathcal{L}(\sum_{i=1}^{n-1} X_i(\mathbf{x}) \cdot T_i(t) \cdot K_i(k))$.

The weak form (41) can be solved by using any appropriate discretization technique for elliptic boundary value problems.

When looking for T_n the test function u^* writes

$$u^*(\mathbf{x}, t, k) = X_n \cdot T_n^* \cdot K_n. \tag{42}$$

By introducing Eqs. (42) and (38) into (39) and integrating in $\Omega \times \mathcal{K}$ it results

$$\int_{\mathcal{I}} T_n^* \cdot \left(\alpha^t \cdot \mathcal{L}_t(T_n(t)) + \beta^t \cdot T_n(t) + g_n^t(t)\right) \; dt = 0, \tag{43}$$

where α^t and β^t are two constants and $g_n^t(t)$ is the function that results from the integration in $\Omega \times \mathcal{K}$ of $f(\mathbf{x}, t)$ and $\mathcal{L}(\sum_{i=1}^{n-1} X_i(\mathbf{x}) \cdot T_i(t) \cdot K_i(k))$.

This equation can be solved by using any appropriate discretization technique for initial value problems.

Finally, when looking for K_n the test function u^* writes

$$u^*(\mathbf{x}, t, k) = X_n \cdot T_n \cdot K_n^*. \tag{44}$$

By introducing Eqs. (44) and (38) into (39) and integrating in $\Omega \times \mathcal{I}$ it results

$$\int_{\mathcal{K}} K_n^* \cdot \left(\alpha^k \cdot K_n(k) + \beta^k \cdot k \cdot K_n(k) + g_n^k(k)\right) \, dk = 0, \tag{45}$$

where α^k and β^k are two constants and $g_n^k(k)$ is the function that results from the integration in $\Omega \times \mathcal{I}$ of $f(\mathbf{x}, t)$ and $\mathcal{L}(\sum_{i=1}^{n-1} X_i(\mathbf{x}) \cdot T_i(t) \cdot K_i(k))$.

Eq. (45) represents an algebraic problem, which is hardly a surprise since the original equation (36) does not contain derivatives with respect to the parameter k. Introduction of the parameter k as additional model coordinate does not increase the cost of a particular enrichment step. It does however necessitate more enrichment steps, i.e. more terms (higher N) in the decomposition (38).

In this case we can conclude that the complexity of the PGD procedure to compute the approximation (38) is of some tens of 3D steady-state problems (the cost related to the 1D IVP and algebraic problems being negligible with respect to the 3D problems). In a classical approach, one must solve for each particular value of the parameter k a 3D problem at each time step. In usual applications, this often implies the computation of several millions of 3D solutions. Clearly, the CPU time savings by applying the PGD can be of several orders of magnitude.

2.5 Discussion on the coordinates separation

When performing a separated representation of the solution of a problem described by a PDE the first step relies in the choice of the coordinates to be separated. We could consider the following scenarios:

- Efficient solvers for transient problems can be defined by applying a space-time separation:

$$u(\mathbf{x}, t) \approx \sum_{i=1}^{N} X_i(\mathbf{x}) \cdot T_i(t). \tag{46}$$

  The constructor of that separated representation was illustrated in the previous section. We cited previously the pioneering works of Ladeveze's team in the field of structural mechanics. Space-time separated representations were also considered in the context of the multiscale coupling of diffusion and kinetic models endowed with very different characteristic times (Chinesta et al. (2010)) as well as in the context of time multi-scale models (Ammar, Chinesta et al. (2012)).
- The fully three-dimensional solution of models defined in degenerate domains is also an appealing field of application of the PGD. Consider

the unknown field $u(x, y, z)$ defined in a domain Ω. Two approaches come to mind:

- Fully separated decomposition.

$$u(x, y, z) \approx \sum_{i=1}^{N} X_i(x) \cdot Y_i(y) \cdot Z_i(z). \tag{47}$$

  This strategy is particularly suitable for fully separable domains, i.e. $\Omega = \Omega_x \times \Omega_y \times \Omega_z$ (Chinesta et al. (2008); Lamari et al. (2010); Pruliere et al. (2013)). For general domains, embedding Ω into a larger separable domain $\tilde{\Omega} = \tilde{\Omega}_x \times \tilde{\Omega}_y \times \tilde{\Omega}_z$ can also be done, as described in Gonzalez et al. (2010).
- 2D-1D decompositions are suitable in plate, shell or profiled geometries. In these cases the most natural decomposition reads

$$u(x, y, z) \approx \sum_{i=1}^{N} X_i(x, y) \cdot Z_i(z). \tag{48}$$

  In plate-type domains $z \in \mathcal{I}$, where $\mathcal{I}$ denotes the plate thickness dimension whereas in extruded profiles it denotes the extrusion direction being Ξ the transversal section.

- Finally, for applications requiring many solutions of a particular model, it suffices to introduce all sources of variability as extra-coordinates. The solution of the resulting parametric multidimensional model is then sought in the separated form

$$u(\mathbf{x}, t, p_1, \cdots p_Q) \approx \sum_{i=1}^{N} X_i(\mathbf{x}) \cdot T_i(t) \cdot P_i^1(p_1) \cdots P_i^Q(p_Q), \tag{49}$$

  where the p_i denote the different problem parameters such as material parameters, boundary conditions, applied loads, initial conditions, and/or geometrical parameters (see the recent reviews Chinesta et al. (2010), (2011), (2013) and the references therein).

3 Advanced topics

3.1 Nonlinear models

Robust strategies exist for addressing efficiently history-dependent nonlinearities, making use of the natural separation of the local and non-linear thermo-mechanical behaviour from the global and linear equilibrium in which a space-time separated representation constitutes an essential ingredient allowing spectacular computing time savings (see the Ladeveze's works).

In the context of separated representations standard linearizations were addressed in some of our former works (Ammar, Normandin, Daim et al. (2010) and Pruliere, Ferec et al. (2010)). We start by considering such procedures in order to point out their inherent limitations before considering alternative approaches based on the use of the asymptotic numerical techniques and interpolation strategies recently considered.

Standard linearization strategies

For describing how standard linearization proceed we consider the simple model given by:

$$\begin{cases} \dfrac{\partial u}{\partial t} - k\Delta u & = & u^2 + f(\mathbf{x},t) & \text{in } \Omega \times (0, T_{max}] \\ u(\mathbf{x},t) & = & 0 & \text{on } \partial\Omega \times (0, T_{max}] \\ u(\mathbf{x},0) & = & 0 & \text{in } \Omega \end{cases} \tag{50}$$

where $\Omega \subset \mathcal{R}^D, D \geq 1$, $T_{max} > 0$ and $k > 0$ is the diffusion coefficient. To build-up the approximated solution of (50) by using a separated representation, we considered two approaches: (i) an incremental linearization and (ii) a Newton linearization, both described below.

- *Incremental linearization.* We write the solution of problem (50) in the separated form

$$u(\mathbf{x},t) \approx \sum_{i=1}^{N} X_i(\mathbf{x}) \cdot T_i(t)$$

We suppose that at iteration n, with $n < N$, the n first modes (X_i, T_i), $i = 1, \cdots n$, are already known and that at present iteration we search the new enrichment functional product $R(\mathbf{x}) \cdot S(t)$ such that the updated approximation writes

$$u^{n+1}(\mathbf{x},t) = u^n(\mathbf{x},t) + R(\mathbf{x}) \cdot S(t) = \sum_{i=1}^{n} X_i(\mathbf{x}) \cdot T_i(t) + R(\mathbf{x}) \cdot S(t) \tag{51}$$

The simplest possibility for accounting for the non linearity induced by the term u^2 consists in computing this term at the previous iteration, that is, assuming at the present iteration the following approximation of the non-linear term

$$u^2 \approx \left(\sum_{i=1}^{n} X_i(\mathbf{x}) \cdot T_i(t) \right)^2 \tag{52}$$

- *Newton linearization.* From the solution at iteration n, u^n

$$u^n(\mathbf{x},t) = \sum_{i=1}^{n} X_i(\mathbf{x}) \cdot T_i(t) \tag{53}$$

the solution at the next iteration can be written as $u^{n+1} = u^n + \widetilde{u}$ where $\widetilde{u}$ is the solution of the linearized problem

$$\frac{\partial \widetilde{u}}{\partial t} - k\Delta\widetilde{u} - 2\,u^n\,\widetilde{u} = -\mathcal{R}(u^n) \tag{54}$$

where the residual $\mathcal{R}(u^n)$ reads

$$\mathcal{R}(u^n) \equiv \left(\frac{\partial u^n}{\partial t} - k\Delta u^n - (u^n)^2 - f(\mathbf{x},t)\right) \tag{55}$$

Now, we assume $\widetilde{u}(\mathbf{x},t) = R(\mathbf{x}) \cdot S(t)$ and $\widetilde{u}^\star = S \cdot R^\star + R \cdot S^\star$. To compute both functions $R(\mathbf{x})$ and $S(t)$ we apply again the alternating directions method deeply described in the previous sections.

- *Numerical results.* We consider the problem (50) where $D = 1$, $\Omega = (0,1)$, $T_{max} = 1$, $f = (16k\pi^2 t + 1)\sin(4\pi x) - t^2\sin^2(4\pi x)$, $\Delta t = 10^{-3}$, $\Delta x = 10^{-3}$ and $k = 1$, whose exact solution reads $u_{ref} = t\sin(4\pi x)$.

When we applied any of the just described strategies, the convergence was reached in 6 iterations, that is, only 6 couples or functions were needed to approximate the solution at the prescribed precision. However, this problem deserves some additional comments. As just indicated the solution of this problem can be written as a single product of functions. At each iteration of the non-linear solver a new product of functions is added. Thus, as the non-linear solver converged in 6 iterations, 6 couples of functions were computed. Obviously, after convergence, if we apply the singular value decomposition to the computed solution (that consists of the sum of the 6 products of functions generated by the non-linear solver) the number of products is reduced to one.

It is clear that in all cases the separated representation algorithm is not optimal in the sense that even if the exact solution can be written from a certain number of functional products, the computed solution could involve a higher number of functional products because the approximation errors and the strategy used for computing the enrichment functions. However, in the case of non-linear models these effects are more noticeable because the number of functional products also depends strongly on the linearization scheme considered.

If instead of considering both linearization schemes just described, one considers the incremental linearization where the non-linear term is assumed given at the previous iteration of the alternating directions non-linear solver, that is, replacing Eq. (52) by

$$u^2 \approx \left(\sum_{i=1}^{n} X_i(\mathbf{x}) \cdot T_i(t) + \overline{S}(\mathbf{x}) \cdot \overline{R}(t) \right)^2 \tag{56}$$

where $\overline{S}(\mathbf{x})$ and $\overline{R}(\mathbf{t})$ are the just computed enrichment functions at the previous iteration of the non-linear enrichment solver, and fixing the same level of precision in the solution, only one product of functions results after convergence. Despite this reduction on the number of sums in the separated representation, the computing time is similar to the one associated to the other strategies.

- *Discussion.* The main difficulty related to the use of standard linearizations lies in the necessity of evaluating the non-linear term. Because the necessity of arriving to a separated representation of such term, one possibility consists in performing a polynomial expansion and then compute the different powers of the solution at the previous step. However, it is easy to understand that the power p of u^n involves too many terms when p, n or D (the dimension of the model) increase as can be noticed from:

$$(u^n)^p = \left(\sum_{i=1}^{i=n} F_i^1(x_1) \times \cdots \times F_i^D(x_D) \right)^p \tag{57}$$

  The issue related to the high powers could be alleviated by introducing new variables v^i such that $v^{(2)} = u \cdot u$, $v^{(3)} = u^3 = v^{(2)} \cdot u$, ... that allows never compute more than powers of degree 2.

  However, in general the differential operators related to the searched functional product involve the more and more terms. Thus, when D, p or N increases the solution procedure described in the previous sections become rapidly inefficient. Next sections explore the use of an alternative approaches, the asymptotic numerical method and interpolation-based strategies.

The asymptotic numerical method

In this section we propose an alternative approach, within the asymptotic numerical method – ANM –framework (Cochelin et al. (1994)) able to include the parametric dimensions as extra-coordinates without modifying the differential operator representation.

We consider again the model defined by

$$\frac{\partial u}{\partial t} - k\Delta u = u^2 + f(\mathbf{x}, t) \tag{58}$$

The non linear term can be affected by a "loading" parameter λ

$$\frac{\partial u}{\partial t} - k\Delta u = \lambda \cdot u^2 + f(\mathbf{x}, t) \tag{59}$$

We denote by u_0 the solution related to $\lambda = \lambda_0 = 0$ that can be computed easily because it corresponds to the solution of the linear problem. The searched solution is the one related to $\lambda = 1$.

Now, we define an asymptotic expansion of the unknown field u as well as of the loading parameter λ by considering the powers of a:

$$\begin{cases} u = u_0 + a \cdot u_1 + a^2 \cdot u_2 \ + \cdots \\ \lambda = \lambda_0 + a \cdot \lambda_1 + a^2 \cdot \lambda_2 \ + \cdots \end{cases} \tag{60}$$

The non linear term can be written as:

$$u^2 = (u^2)_0 + a \cdot (u^2)_1 + a^2 \cdot (u^2)_2 \ + \cdots + a^p \cdot (u^2)_p + \cdots \tag{61}$$

where $(u^2)_p$ reads:

$$(u^2)_p = \sum_{i=0}^{i=p} u_i \cdot u_{p-i} = 2 \cdot u_0 \cdot u_p + \sum_{i=1}^{i=p-1} u_i \cdot u_{p-i} \tag{62}$$

Introducing (60) and (62) into (59) and identifying the different powers of a, it results:

- Power $p = 0$. Solution u_0 is assumed known because it corresponds to the solution of the linear problem resulting from $\lambda = \lambda_0 = 0$.
- Power $p = 1$. We should solve:

$$\frac{\partial u_1}{\partial t} - k\Delta u_1 = \lambda_0 \cdot (u^2)_1 + \lambda_1 \cdot (u^2)_0 \tag{63}$$

and because $\lambda_0 = 0$, the evolution equation for u_1 reduces to:

$$\frac{\partial u_1}{\partial t} - k\Delta u_1 = \lambda_1 \cdot (u^2)_0 \tag{64}$$

We can notice that the previous equation contains two unknowns, u_1 and λ_1. The solution procedure needs an extra-equation. The following equation is considered (Cochelin et al. (1994)):

$$(u - u_0, u_1) + (\lambda - \lambda_0, \lambda_1) = a \tag{65}$$

where (,) denotes the scalar product. The first order of (65) reads

$$(u_1, u_1) + (\lambda_1)^2 = 1 \tag{66}$$

Because the linearity of Eq. (64) if $\hat{u}_1$ is the solution of

$$\frac{\partial \hat{u}_1}{\partial t} - k\Delta\hat{u}_1 = (u^2)_0 \tag{67}$$

the searched solution u_1 results:

$$u_1 = \lambda_1 \cdot \hat{u}_1 \tag{68}$$

By introducing Eq. (68) into Eq. (66) it results:

$$(\lambda_1)^2 \cdot [(\hat{u}_1, \hat{u}_1) + 1] = 1 \tag{69}$$

that allows computing the value of λ_1 and then u_1 from Eq. (68).

- Power p, with $p > 1$. We should solve:

$$\frac{\partial u_p}{\partial t} - k\Delta u_p = \lambda_p \cdot (u^2)_0 + \sum_{i=1}^{i=p-1} \lambda_i \cdot (u^2)_{p-i} \tag{70}$$

where the fact that $\lambda_0 = 0$ was taken into account.

In this expression we can notice that the differential operator is always the same and that in the right hand member the term $\sum_{i=1}^{i=p-1} \lambda_i \cdot (u^2)_{p-i}$ has been already computed.

Eq. (65) at order p reads:

$$(u_p, u_1) + (\lambda_p, \lambda_1) = 0 \tag{71}$$

Because the linearity of Eq. (70) if $\tilde{u}_p$ is the solution of

$$\frac{\partial \tilde{u}_p}{\partial t} - k\Delta\tilde{u}_p = \sum_{i=1}^{i=p-1} \lambda_i \cdot (u^2)_{p-i} \tag{72}$$

and $\hat{u}_p$ the solution of

$$\frac{\partial \hat{u}_p}{\partial t} - k\Delta\hat{u}_p = (u^2)_0 \tag{73}$$

that as we can notice corresponds to $\hat{u}_1$, the searched solution u_p reads:

$$u_p = \tilde{u}_p + \lambda_p \cdot \hat{u}_1 \tag{74}$$

By introducing Eq. (74) into Eq. (71) it results:

$$(\tilde{u}_p, \hat{u}_1) + \lambda_p \cdot [(\hat{u}_1, \hat{u}_1) + 1] = 0 \tag{75}$$

that allows computing the value of λ_p and then u_p from Eq. (74).

Numerical results. Since we have seen that the PGD offers a powerful alternative to classical finite elements when one can take advantage of variable separation, we next investigate the use of the PGD as an alternative of finite elements in the solution of a non-linear heat transfer problem within the ANM framework. For this purpose we condsider:

$$-k\Delta u - f + u^2 = 0. \tag{76}$$

Here $u(x, y) \in \Omega_x \times \Omega_y$, with $\Omega_x = \Omega_y = (0, 2)$, the source term being constant, $f = 1$ and the conductivity k is also constant and equal to 1. We consider the following non-homogeneous Dirichlet conditions along the boundary Γ of Ω: $u(x, y) = 2$ for $(x, y) \in \Gamma$. For this choice of boundary conditions, it is obvious that the non linear term will greatly influence the solution. Indeed, accounting or not for the nonlinearity could change the concavity of the solution.

The PGD solution is sought in the following separated form:

$$u(x, y) \approx \sum_{i=1}^{N} X_i(x) \cdot Y_i(y), \tag{77}$$

where we use a Galerkin finite elements method with continuous piecewise linear interpolation to solve the 1D BVP problems that appear in the PGD algorithm. In order to apply the Asymptotic Numerical Method, we introduce the loading parameter λ and modify the original problem in the following way:

$$-k\Delta u - f + \lambda u^2 = 0. \tag{78}$$

Now, the procedure described in the previous section is applied. In Figure 1 we show the value of the solution in the centre of the domain $u(1, 1)$ for different values of the loading parameter λ and for different orders of expansion. The same problem solved with 2D finite elements on an equivalent mesh instead of the PGD predicts curves that exactly superimpose on the curves depicted in Figure 1. We see that when the loading parameter is zero, the solution is greater than 2 while it becomes lower than 2 for increasing values of λ. Note that the solution of the original non-linear problem corresponds to $\lambda = 1$. Furthermore, we see that the expansion up to order 4 is sufficient for capturing the nonlinearity of the problem.

In this specific example the use of PGD-based ANM has not any advantage since the problem is very simple. There is no real benefit in seeking a separated representation of the solutions of the various linear problems that have to be solved to compute the solution at the different orders within the ANM framework. Indeed as all the linear problems appearing in the ANM

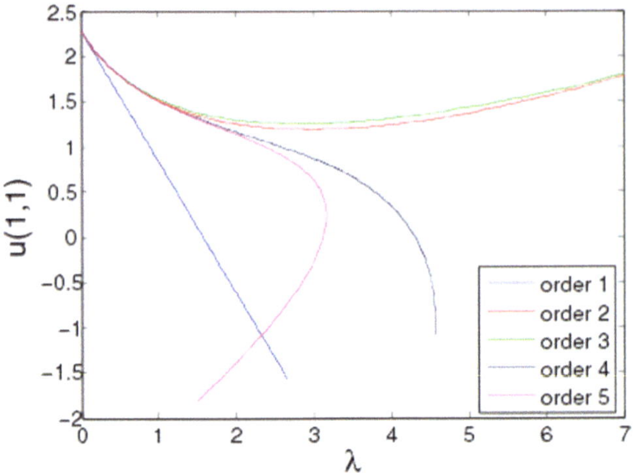

Figure 1. Convergence of the ANM at the centre of the domain for different orders of expansion of the loading parameter λ.

at its different orders involve the same differential operator, the computational cost of using 2D finite elements is dominated by the initial construction and factorization of the matrix representing the discretized operator. The successive solutions are then inexpensive. With the PGD, one cannot take advantage of this fact because the coefficient related to the differential operators are changing at each iteration.

We now turn to the case of parametric models, for which the PGD is well suited. We solve the same non linear problem but consider this time the diffusivity k as an additional coordinate:

$$-k\Delta u - 1 + u^2 = 0. \tag{79}$$

Here $u(x, y, k) \in \Omega_x \times \Omega_y \times \mathcal{K}$, with $\Omega_x = \Omega_y = (0, 2)$ and $k \in \mathcal{K} = [1, 5]$. We consider the same Dirichlet condition as previously and the same expansion for the loading parameter. Applying the ANM directly to such model is only possible due to the PGD ability to treat multidimensional parametric models.

The interested read can refer to Leygue et al. (2013) for an exhaustive analysis on the application of ANM-based PGD for solving steady state, transient and parametric nonlinear models. Despite the noticed advantages when considering this kind of linearization it seems difficult to consider it in the case of multi-parametric models.

A promising alternative consists of interpolating the non-linear term as proposed in Barrault et al. (2004), whose discrete counterpart, the Discrete Empirical Interpolation Method – DEIM – was proposed in Chaturantabut and Sorensen (2010), and then introducing the interpolated non-linear term within the separated representation PGD constructor.

DEIM based PGD

We consider the solution of

$$\frac{\partial u}{\partial t} - k \cdot \Delta u = \mathcal{L}(u) + f(\mathbf{x}, t) \tag{80}$$

with homogeneous initial and boundary conditions and where $\mathcal{L}(u)$ represents a nonlinear function of u.

We first address the solution $u^0(\mathbf{x}, t)$ of its linear counterpart:

$$\frac{\partial u^0}{\partial t} - k \cdot \Delta u^0 = f(\mathbf{x}, t) \tag{81}$$

whose solution is found in a separated form by applying the PGD. Thus, the solution of Eq. (81) writes:

$$u^0(\mathbf{x}, t) \approx \sum_{i=1}^{i=N^0} X_i^0(\mathbf{x}) \cdot T_i^0(t) \tag{82}$$

that allows to define the reduced approximation basis $\mathcal{B}^0 = \{\tilde{X}_1^0 \cdot \tilde{T}_1^0, \cdots, \tilde{X}_{N^0}^0 \cdot \tilde{T}_{N^0}^0\}$ that contains the normalized functions: $\tilde{X}_i^0 = \frac{X_i^0}{\|X_i^0\|}$ and $\tilde{T}_i^0 = \frac{T_i^0}{\|T_i^0\|}$.

Now, we could define an interpolation of the nonlinear function $\mathcal{L}(u)$ by using the basis $\mathcal{B}^0$. For this purpose we consider, as proposed in Chinesta et al. (2013), N^0 points $(\mathbf{x}_j^0, t_j^0)$, $j = 1, \cdots, N^0$, and we enforce:

$$\mathcal{L}(u^0(\mathbf{x}_j^0, t_j^0)) = \sum_{i=1}^{i=N^0} \xi_i^0 \cdot \tilde{X}_i^0(\mathbf{x}_j^0) \cdot \tilde{T}_i^0(t_j^0), \quad j = 1, \cdots, N^0 \tag{83}$$

that represents a linear system of size N^0 whose solution allows calculating the coefficients ξ_i^0.

As soon as these coefficients ξ_i^0 are known, the interpolation of the non-linear term is fully defined:

$$\mathcal{L}^0 \equiv \mathcal{L}(u^0(\mathbf{x}, t)) \approx \sum_{i=1}^{i=N^0} \xi_i^0 \cdot \tilde{X}_i^0(\mathbf{x}) \cdot \tilde{T}_i^0(t) \tag{84}$$

that is introduced into the original nonlinear problem leading to the linear problem involving now $u^1(\mathbf{x}, t)$

$$\frac{\partial u^1}{\partial t} - k \cdot \Delta u^1 = \mathcal{L}^0(u) + f(\mathbf{x}, t) \tag{85}$$

Now for calculating the solution $u^1(\mathbf{x}, t)$ many choices exist, being the most direct ones, among many others:

- Restart the separated representation, i.e.:

$$u^1(\mathbf{x}, t) \approx \sum_{i=1}^{i=N_s^1} X_i^1(\mathbf{x}) \cdot T_i^1(t) \tag{86}$$

- Reuse the solution u^0:

$$u^1(\mathbf{x}, t) \approx u^0(\mathbf{x}, t) + \sum_{i=1}^{i=N_r} X_i^1(\mathbf{x}) \cdot T_i^1(t) = \sum_{i=1}^{i=N_r^1} X_i^1(\mathbf{x}) \cdot T_i^1(t) \tag{87}$$

- Reuse by projecting. In this case first we consider

$$u^{1,0}(\mathbf{x}, t) \approx \sum_{i=1}^{i=N^0} \eta_i \cdot \tilde{X}_i^0(\mathbf{x}) \cdot \tilde{T}_i^0(t) \tag{88}$$

  that introduced into (85) allows computing the coefficients η_i. Then the approximation is enriched by considering

$$u^1(\mathbf{x}, t) \approx \sum_{i=1}^{i=N^0} \eta_i \cdot \tilde{X}_i^0(\mathbf{x}) \cdot \tilde{T}_i^0(t) + \sum_{i=1}^{i=N_p} X_i^1(\mathbf{x}) \cdot T_i^1(t) = \sum_{i=1}^{i=N_p^1} X_i^1(\mathbf{x}) \cdot T_i^1(t) \tag{89}$$

We define both, N^1 as the number of final approximation functions, N_s^1, N_r^1 or N_p^1 depending on the previous choice, and the associated reduced approximation basis $\mathcal{B}^1 = \{\tilde{X}_1^1 \cdot \tilde{T}_1^1, \cdots, \tilde{X}_{N^1}^1 \cdot \tilde{T}_{N^1}^1\}$. Now the nonlinear term is interpolated again from N^1 points $(\mathbf{x}_j^1, t_j^1)$, $j = 1, \cdots, N^1$:

$$\mathcal{L}^1 \equiv \mathcal{L}(u^1(\mathbf{x}, t)) \approx \sum_{i=1}^{i=N^1} \xi_i^1 \cdot \tilde{X}_i^1(\mathbf{x}) \cdot \tilde{T}_i^1(t) \tag{90}$$

that is introduced into the original nonlinear problem leading to the linear problem involving now $u^2(\mathbf{x}, t)$. The just described procedure is repeated until reaching convergence.

The only point that deserves additional comments is the one related to the choice of the interpolation points $(\mathbf{x}_j^k, t_j^k)$, $j = 1, \cdots, N^k$ at iteration k. At this iteration the reduced approximation basis reads:

$$\mathcal{B}^k = \{\tilde{X}_1^k \cdot \tilde{T}_1^k, \cdots, \tilde{X}_{N^k}^k \cdot \tilde{T}_{N^k}^k\} \tag{91}$$

Following Barrault et al. (2004) and Chaturantabut and Sorensen (2010) we consider

$$(\mathbf{x}_1^k, t_1^k) = argmax_{\mathbf{x},t} |\tilde{X}_1^k(\mathbf{x}) \cdot \tilde{T}_1^k(t)| \tag{92}$$

then we compute d_1 from

$$d_1 \cdot \tilde{X}_1^k(\mathbf{x}_1^k) \cdot \tilde{T}_1^k(t_1^k) = \tilde{X}_2^k(\mathbf{x}_1^k) \cdot \tilde{T}_2^k(t_1^k) \tag{93}$$

that allows defining $r_2^k(\mathbf{x}, t)$

$$r_2^k(\mathbf{x}, t) = \tilde{X}_2^k(\mathbf{x}) \cdot \tilde{T}_2^k(t) - d_1 \cdot \tilde{X}_1^k(\mathbf{x}) \cdot \tilde{T}_1^k(t) \tag{94}$$

from which computing point $(\mathbf{x}_2^k, t_2^k)$ according to:

$$(\mathbf{x}_2^k, t_2^k) = argmax_{\mathbf{x},t} |r_2^k(\mathbf{x}, t)| \tag{95}$$

As by construction $r_2^k(\mathbf{x}_1^k, t_1^k) = 0$ we can ensure $(\mathbf{x}_2^k, t_2^k) \neq (\mathbf{x}_1^k, t_1^k)$.

The procedure is generalized for obtaining the other points involved in the interpolation procedure. Thus for obtaining point $(\mathbf{x}_j^k, t_j^k)$, $j \leq k$, we consider

$$r_j^k(\mathbf{x}, t) = \tilde{X}_j^k(\mathbf{x}) \cdot \tilde{T}_j^k(t) - \sum_{i=1}^{i=j-1} d_i \cdot \tilde{X}_i^k(\mathbf{x}) \cdot \tilde{T}_i^k(t) \tag{96}$$

whose maximum results the searched point $(\mathbf{x}_j^k, t_j^k)$, i.e.

$$(\mathbf{x}_j^k, t_j^k) = argmax_{\mathbf{x},t} |r_j^k(\mathbf{x}, t)| \tag{97}$$

The coefficients $d_1, \cdots, d_{j-1}$ must be chosen for ensuring that $(\mathbf{x}_j^k, t_j^k) \neq (\mathbf{x}_i^k, t_i^k)$, $\forall i < j \leq k$. For this purpose we enforce that the residual $r_j^k(\mathbf{x}, t)$ vanishes at each location $(\mathbf{x}_i^k, t_i^k)$ with $i < j$, that is:

$$r_j^k(\mathbf{x}_l^k, t_l^k) = 0 = \tilde{X}_j^k(\mathbf{x}_l^k) \cdot \tilde{T}_j^k(t_l^k) - \sum_{i=1}^{i=j-1} d_i \cdot \tilde{X}_i^k(\mathbf{x}_l^k) \cdot \tilde{T}_i^k(t_l^k), \; l = 1, \cdots, j-1 \tag{98}$$

that constitutes a linear system whose solution results the searched coefficients $d_1, \cdots, d_{j-1}$.

Consider the one-dimensional model

$$\frac{\partial u}{\partial t} - k \cdot \frac{\partial^2 u}{\partial x^2} = -u^2 + f(x,t), \quad (x,t) \in (0,1) \times (0,1] \tag{99}$$

where the source term $f(x,t)$ is chosen in order to ensure that the exact solution writes

$$u(x,t) = x \cdot t + x^2 \cdot t^2 \tag{100}$$

that represents a separated solution involving two terms.

With this choice the initial condition reads $u(x, t = 0) = 0$ whereas the boundary conditions are given by $u(x = 0, t) = 0$ and $\frac{\partial u}{\partial x}(x = 1, t) = t + 2 \cdot t^2$. The strategy that reuses the previous reduced bases (Eq. (87)), the convergence was reached after the construction of 5 reduced bases ($k = 5$) in which the nonlinear term was interpolated. The final solution involved 40 ($N^k = 40$) functional products $X_i^k(x) \cdot T_i^k(t)$, $i = 1, \cdots, 40$. Figures 2 and 3 depict the six first space and time modes respectively. Then Fig. 4 compares the time evolution at different locations obtained with the DEIM based PGD and the exact solution. Finally Fig. 5 shows the space-time DEIM based PGD solution. From these results we can conclude on the potentiality of the proposed technology for solving non-linear eventually multi-parametric models, as illustrated in Chinesta et al. (2013).

3.2 Advective stabilization

It is well-known that standard finite element (Galerkin) methods are not suitable for convection-dominated problems, since they yield unstable, spuriously oscillating solutions (Donea and Huerta (2002))

Among the numerous stabilization schemes that have been proposed for convection-dominated equations, the Streamline-Upwind/Petrov-Galerkin (SUPG) method (Hughes (1995)) is one of the most popular. When a reaction term is important, Sub-Grid Scale (SGS) techniques have been advocated to reduce spurious oscillations (Hughes et al. (1998)). An inherent difficulty of all these methods is the choice of the stabilization parameter. In fact, algebraic or asymptotic analyses have been developed for one-dimensional problems, but optimal stabilization parameters are not easily obtained in higher dimensions.

The PGD is well suited in this context. Indeed, as it leads at each enrichment step to the solution of a large number of decoupled one-dimensional problems, the PGD allows for the selection of optimal stabilization param-

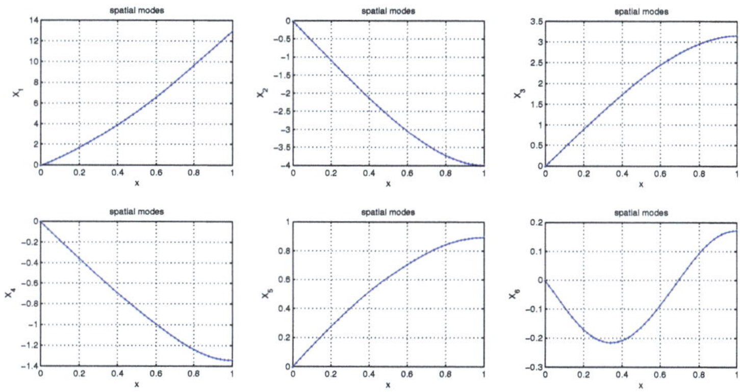

Figure 2. Space modes: $X_i^k(x),\ i = 1, \cdots, 6$

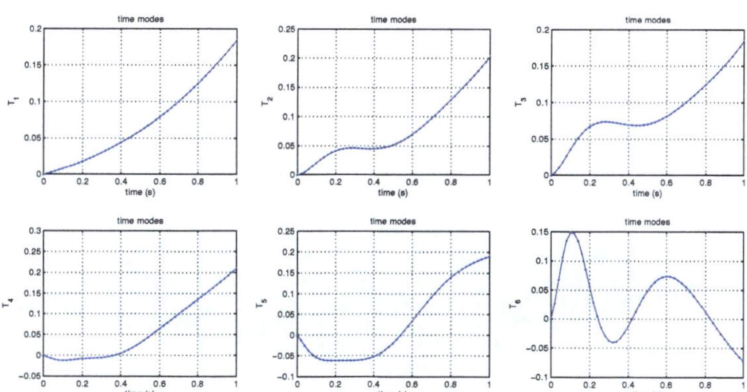

Figure 3. Time modes: $T_i^k(t),\ i = 1, \cdots, 6$

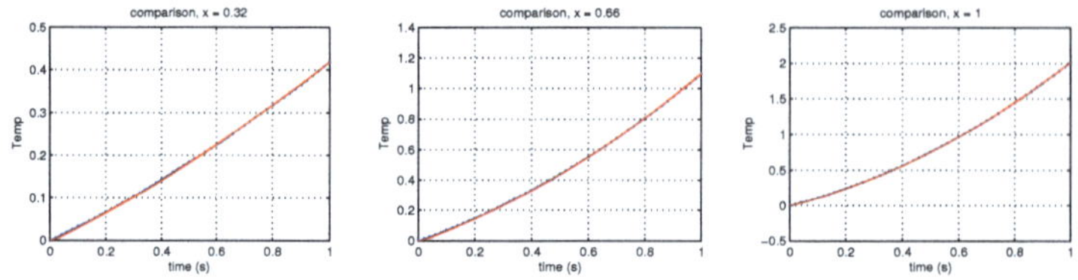

Figure 4. DEIM based PGD solution versus the exact one

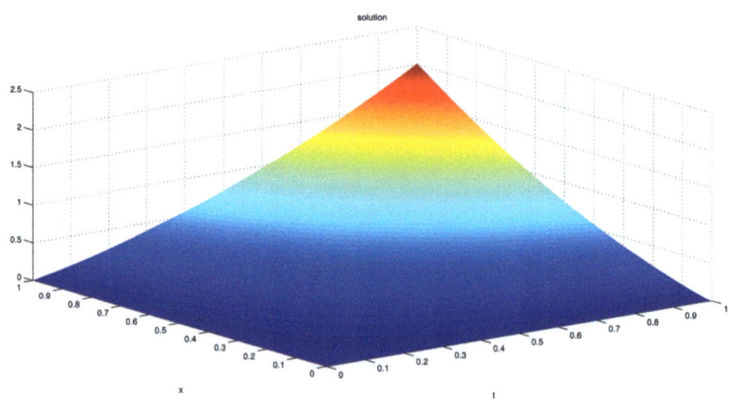

Figure 5. Space-time reconstructed DEIM based PGD solution

eters to account for the convection-dominated character of each of these problems.

Following Gonzalez et al. (2013), we illustrate this point with a two-dimensional convection-diffusion equation defined in a rectangular domain:

$$\mathbf{v} \cdot \nabla u - k \cdot \Delta u = f, \tag{101}$$

with $u(\mathbf{x})$, $\mathbf{x} \in \Omega$, $\Omega = \Omega_x \times \Omega_y = (0, L) \times (0, H)$, and $\mathbf{v}$ constant in Ω. The weighted residual integral form related to Eq. (101) reads

$$\int_{\Omega_x \times \Omega_y} u^* \cdot \left(v_x \cdot \frac{\partial u}{\partial x} + v_y \cdot \frac{\partial u}{\partial y} - k \cdot \frac{\partial^2 u}{\partial x^2} - k \cdot \frac{\partial^2 u}{\partial y^2} - f \right) \, dx \cdot dy \;=\; 0, \tag{102}$$

for all suitable test functions u^*.

The PGD solution is sought in the separated form

$$u(x, y) = \sum_{i=1}^{N} X_i(x) \cdot Y_i(y). \tag{103}$$

At enrichment step n of the PGD algorithm, we have already computed the approximation

$$u^{n-1}(x, y) = \sum_{i=1}^{n-1} X_i(x) \cdot Y_i(y). \tag{104}$$

and we wish to obtain the next one, i.e.

$$u^n(x, y) = u^{n-1}(x, y) + X_n(x) \cdot Y_n(y) = \sum_{i=1}^{n-1} X_i(x) \cdot Y_i(y) + X_n(x) \cdot Y_n(y). \tag{105}$$

An alternating direction iterative scheme is then used to solve the non-linear problem for $X_n(x)$ and $Y_n(y)$. At iteration p, we must compute $X_n^p(x)$ from $Y_n^{p-1}(y)$, and then $Y_n^p(x)$ from $X_n^p(x)$. Let us detail the first step. At this stage, the approximation reads

$$u^n(x, y) = \sum_{i=1}^{n-1} X_i(x) \cdot Y_i(y) + X_n^p(x) \cdot Y_n^{p-1}(y), \tag{106}$$

where all functions except $X_n^p(x)$ are known. Selecting $u^*(x, y) = X_n^*(x) \cdot Y_n^{p-1}(y)$ for the test function and introducing (106) into (102), we obtain

$$\int_{\Omega_x \times \Omega_y} X_n^* \cdot Y_n^{p-1} \cdot \left(v_x \cdot \frac{dX_n^p}{dx} \cdot Y_n^{p-1} + \right.$$

$$\left. + v_y \cdot X_n^p \cdot \frac{dY_n^{p-1}}{dy} - k \cdot \frac{d^2 X_n^p}{dx^2} \cdot Y_n^{p-1} - k \cdot X_n^p \cdot \frac{d^2 Y_n^{p-1}}{dy^2} \right) dx \cdot dy =$$

$$- \int_{\Omega_x \times \Omega_y} X_n^* \cdot Y_n^{p-1} \cdot$$

$$\sum_{i=1}^{n-1} \left(v_x \cdot \frac{dX_i}{dx} \cdot Y_i + v_y \cdot X_i \cdot \frac{dY_i}{dy} - k \cdot \frac{d^2 X_i}{dx^2} \cdot Y_i - k \cdot X_i \cdot \frac{d^2 Y_i}{dy^2} \right) dx \cdot dy +$$

$$\int_{\Omega_x \times \Omega_y} X_n^* \cdot Y_n^{p-1} \cdot f \; dx \cdot dy. \tag{107}$$

Integration over Ω_y then yields the weighted residual form of a one-dimensional problem for the unknown function $X_n^p(x)$, which involves known coefficients α^x, β^x, γ^x, δ_i^x, χ_i^x, υ_i^x and ξ^x whose definition is left as an exercise:

$$\int_{\Omega_x} X_n^* \cdot \left(\alpha^x \cdot \frac{d^2 X_n^p}{dx^2} + \beta^x \cdot \frac{dX_n^p}{dx} + \gamma^x \cdot X_n^p \right) dx =$$

$$- \int_{\Omega_x} X_n^* \cdot \sum_{i=1}^{n-1} \left(\delta_i^x \cdot \frac{d^2 X_i}{dx^2} + \chi_i^x \cdot \frac{dX_i}{dx} + \upsilon_i^x \cdot X_i \right) dx + \int_{\Omega_x} X_n^* \cdot \xi^x \; dx. \tag{108}$$

The corresponding strong form is a one-dimensional convection-diffusion-reaction equation with a source term, for which quasi-optimal stabilization methods exist (Donea and Huerta (2002)). The interested reader can refer to Gonzalez et al. (2013) for a deeper analysis of this topic and numerical tests proving the performance of this approach, in particular for high-dimensional problems.

3.3 Adaptivity

This section aims at introducing an objective error assessment as a convergence criterion for the iterative scheme successively enriching the separated representation. The main goal is to compute the number of terms in the finite sums decomposition such that the error related to a certain quantity of interest is lower than a certain threshold value. The error assessment is performed on the basis of an arbitrary linear quantity of interest represented by a linear functional $o(\cdot)$. Thus, the enrichment scheme loops while the error in this quantity is larger than the prescribed tolerance and stops once the prescribed accuracy is reached.

The error in the quantity of interest is represented using an auxiliary problem (denoted as dual or adjoint problem). This strategy has been widely used by many authors in the field of error estimation and adaptivity, using different error estimation techniques (see Ammar, Chinesta et al. (2010) and the references therein).

Here, in order to obtain a proper error assessment, the exact solution of the auxiliary problem (also denoted as extractor) is simply replaced by a more accurate solution. In the context of PGD, this option is very effective and costless because it only requires performing some extra enrichment iterations for the adjoint problem.

In this section we are assuming a generic multidimensional model whose weak form writes

$$a(u, u^*) = b(u^*) \tag{109}$$

defined in $\Omega = \Omega_1 \times \cdots \times \Omega_D$, where each Ω_d $(d = 1, 2, \cdots, D)$ involves the coordinate x_d (not necessarily one-dimensional). From now on, the form (109) will be referred as primal form.

The solution of this weak form by assuming the separated representation of the unknown field $u(x_1, \cdots, x_D)$

$$u \approx \sum_{j=1}^{i=N} F_i^1(x_1) \times \cdots \times F_i^D(x_D) \tag{110}$$

has been deeply described previously.

Now, we are interested in a certain function of u, $o(u)$, of physical interest (the model output). In what follows we assume that the operator defining the output is linear. Thus, we could write

$$o(u) = \mathcal{O}\left(\sum_{j=1}^{i=N} F_i^1(x_1) \times \cdots \times F_i^D(x_D)\right) \tag{111}$$

If this operator accepts a separated representation, the error on the output can be evaluated by solving the so-called dual problem:

$$a(u^*, v) = o(u^*) \tag{112}$$

whose solution v can be computed in a separated form within the PGD framework.

A good error estimation needs for an accurate solution of the dual problem. Within the separated representation framework the solution of the dual problem (112) (assumed accurate enough) can be written as:

$$v \approx \sum_{j=1}^{j=N_d} G_i^1(x_1) \times \cdots \times G_i^D(x_D) \tag{113}$$

Now, the error in the output can be evaluated from:

$$o(e) = b(v) - a(u,v) \tag{114}$$

Accounting for local behaviors. On the other hand coupling standard mesh based discretization techniques with reduced bases (POD and PGD) is an issue of major interest in order to represent localized behaviors (e.g. discontinuities, singularities, boundary layers, ...). Some coupling strategies were proposed in Ammar, Pruliere, Ferec et al. (2009) and in Ammar et al. (2011)

3.4 Coupled models

We are considering a simple parabolic model in a one-dimensional physical space Ω:

$$\frac{\partial u}{\partial t} - a\Delta u = f(x,t) \qquad \text{in } \Omega \times (0, T_{max}] \tag{115}$$

with the following initial and boundary conditions

$$\begin{cases} u(x,0) = u^0 & x \in \Omega, \\ u(x,t) = u_g & (x,t) \in \partial\Omega \times (0, T_{max}] \end{cases} \tag{116}$$

We are assuming that the term source depends on the local value of r fields $C_i(x,t)$, $i = 1, \cdots, r$:

$$f(x,t) = \sum_{i=1}^{i=r} \gamma_i \cdot C_i(x,t) \tag{117}$$

where the time evolution of the r fields $C_i(x,t)$ is governed by r coupled ordinary differential equations (the so-called kinetic model). For the sake of simplicity we consider the linear case, the non-linear one reduces to a sequence of linear problems by applying an appropriate linearization strategy. The system of linear ODEs writes at each point $x \in \Omega$:

$$\frac{dC_i(x,t)}{dt} = \sum_{j=1}^{j=r} \alpha_{ij}(x)\; C_j(x,t) \tag{118}$$

We are assuming that the kinetic coefficients α_{ij} evolves smoothly in Ω, because in practical applications these coefficients depend on the solution of the diffusion problem, $u(x,t)$. For the sake of simplicity and without loss of generality, the alpha coefficients were assumed in Chinesta et al. (2010) evolving linearly in x, but in the description that follows we assume those coefficients constant.

Now, we are describing three possible procedures for solving Eqs. (115) and (118) deeply analyzed in Chinesta et al. (2010).

1. The simplest strategy consists in using a separated representation of the global problem solution (115) whereas the local problems are integrated in the whole time interval at each nodal position (or integration point). Obviously, this strategy implies the solution of r coupled ordinary differential equations at each node (or integration point). Moreover, the resulting fields $C_i(x,t)$, $r = 1, \cdots, r$, don't have a separated structure, and by this reason before injecting these fields into the global problem (115) we should separate them by invoking, for example, the singular value decomposition (SVD) leading to:

$$C_i(x,t) \approx \sum_{q=1}^{q=m} X_q^{C,i}(x) \cdot T_q^{C,i}(t) \tag{119}$$

   As soon as the source term has a separated structure, the procedure described in the previous chapters can be applied again for computing the new trial solution of the global problem that writes:

$$u(\mathbf{x},t) \approx \sum_{i=1}^{N} X_i^u(\mathbf{x}) \cdot T_i^u(t) \tag{120}$$

   Thus, this coupling strategy requires the solution of many local problems (for all the coupled species at all nodal positions or integration points). Moreover, after these solutions (that we recall that could be performed in parallel) a singular value decomposition must be applied in order to separated these solutions prior to inject them in the PGD solver of the global problem (115).
2. The second coupling strategy lies in globalizing the solution of the local problems. Thus, we assume that the field related to each species can be written in a separated form:

$$C_i(x,t) \approx \sum_{q=1}^{q=m} X_q^{C,i}(x) \cdot T_q^{C,i}(t) \tag{121}$$

and now, we apply the standard procedure to build-up the reduced separated approximation, i.e. for constructing all the functions involved in (121). Thus, instead of solving the r coupled ODEs in Eq. (119) at each nodal position (or integration point), we should solve only r higher dimensional coupled models defined in the physical space and time. Obviously, if the number of nodes (or integration points) is important (mainly when 3D physical spaces are considered) the present coupling strategy could offer significant CPU time savings.

This strategy allows computing directly a separated representation, and then, with respect to the previous one, the application of the SVD is avoided. However, if the number of species is high, the computational efforts can become important, because the space-time separated solver must be applied to each species.

3. The third alternative, that in our opinion is the more appealing one for solving models involving many species, as large as one wants, implies the definition of a new variable $C(x,t,c)$, that as we can notice contains an extra coordinate c, of discrete nature, and that takes integer values: $c = 1, \cdots, r$, in such manner that $C(x,t,i) \equiv C_i(x,t)$, $i = 1, \cdots, r$. Thus, we have increased the dimensionality of the problem, but now, only a single problem should be solved, instead one for each species as was the case when using the previous strategy. This increase of the model dimensionality is not dramatic because the separated representation allows circumventing the curse of dimensionality, allowing for fast and accurate solutions of highly multidimensional models. Now, the issue is the derivation of the governing equation for this new variable $C(x,t,c)$ and the separated representation constructor able to define the approximation:

$$C(x,t,c) \approx \sum_{q=1}^{q=S} X_q^C(x) \cdot T_q^C(t) \cdot A_q(c) \tag{122}$$

An exhaustive description and analysis was carried out in Chinesta et al. (2010).

4 Models defined in plate and shell geometries

The solution of 3D models in degenerated geometries in which some characteristic dimensions are much lower than the other ones —e.g. beams, plates or shells— is a tricky issue when using standard mesh-based discretization techniques. Separated representations allow decoupling the meshes used for approximating the solution along each coordinate. Thus, in plate or shell geometries 3D solutions can be obtained from a sequence of 2D and 1D problems allowing a fine and accurate representation of the solution evolution along the thickness coordinate while keeping the computational complexity characteristic of 2D simulations. In this section we revisit the application of such methodology for addressing different physics (thermal models, solid and fluid mechanics and electromagnetic problems) in such degenerated geometries.

4.1 In-plane-out-of-plane separated representation

Plates and shells are very common in nature and thus they inspired engineers that used both from the very beginning of structural mechanics. In general the design of such structural elements requires the calculation of stresses, strains and displacements for the design loads. Strains and stresses are related by the so-called constitutive law, the simplest one being the linear elasticity. Moreover other physics can be encountered in these structural elements as for example thermal processes, electromagnetism or fluid flows. Typically composites manufacturing processes usually involve a resin flowing into the double scale porous media, that polymerize by a thermal, chemical or electromagnetic activation. Welding processes could also involve the use of thermal, mechanical or electromagnetic devices.

Thus, structural or processes design always involve the solution of a set of partial differential equations in the degenerate domain of the plate or the shell with appropriate initial and boundary conditions. These domains are degenerated because one of its characteristic dimensions (the thickness in the present case) is much lower that the other characteristic dimensions. We will understand the consequences of such degeneracy later. When analytical solutions are neither available nor possible because the geometrical or behavior complexities, the solution must be calculated by invoking any of the available numerical techniques (finite elements, finite differences, finite volumes, methods of particles, ...).

In the numerical framework the solution can be only obtained in a discrete number of points, usually called nodes, properly distributed in the domain. From the solution at those points, it can be interpolated at any other point in the domain. In general regular nodal distributions are preferred because they offer the best accuracy. In the case of degenerated plate or shell domains one could expect that if the solution evolves significantly in the thickness direction, a large enough number of nodes must be distributed along the thickness direction to ensure the accurate representation of the field evolution in that direction. In that case, a regular nodal distribution in the whole domain will imply the use of an extremely large number of nodes with the consequent impact on the numerical solution efficiently.

When simple behaviors and domains were considered, plate and shell theories were developed allowing, through the introduction of some hypotheses, reducing the 3D complexity to the 2D related to the problem now formulated by considering the in-plane coordinates. The use of these theories was extended gradually for addressing larger and more complex geometries (anisotropic laminates, ...) and behaviors. These simplified "plate models" exist for most of the physics just referred, but their applicability is in many cases too constrained.

Moreover, as soon as richer physics are involved in the models and the considered geometries differ of those ensuring the validity of the different reduction hypotheses, efficient simulations are compromised. For example in composites manufacturing processes of large parts many reactions and thermal processes inducing significant evolutions on the thermomechanical fields in the thickness occur. These inhomogeneities are at the origin of residual stresses and the associated distortion of the formed parts.

In these circumstances as just indicated the reduction from the 3D model to a 2D simplified one is not obvious, and 3D simulations appear many times as the only valid route for addressing such models, that despite the fact of being defined in degenerated geometries (plate or shell) they seem requiring a fully 3D solution. In order to integrate such calculations (fully 3D and implying an impressive number of degrees of freedom) in usual design procedures, a new efficient (fast and accurate) solution procedure is needed.

A direct consequence of the separated representations involved in the PGD is the the space separation. Thus an in-plane-out-of-plane decomposition was proposed for solving flow problems in laminates (Chinesta, Ammar et al. (2011); Ghnatios et al. (2013)), thermal problems in extruded geometries (Leygue et al. (2012)) and laminates (Chinesta, Leygue et al. (2014)), elasticity problems in plates (Bognet et al. (2012)) and shells geometries (Bognet et al. (2014)). In general the 3D solution was obtained from the solution of a sequence of 2D problems (the ones involving the in-plane coordinates) and 1D problems (the ones involving the coordinate related to the plate thickness).

It is important emphasizing the fact that these approaches are radically different to standard plate and shell approaches. We proposed a 3D solver able to compute the different unknown fields without the necessity of introducing any hypothesis. The most outstanding advantage is that 3D solutions can be obtained with a computational cost characteristic of standard 2D solutions.

In what follows we revisit the in-plane-out-of-plane representation and its application for addressing different physics (thermal, elastic, flow and electromagnetic models) in plate domains, eventually consisting of a laminate composed of several anisotropic plies.

Elastic problem in plate domains

We proposed in Bognet et al. (2012) an in-plane-out-of-plane decomposition of the 3D elastic solution in a plate geometry. The elastic problem was defined in a plate domain $\Xi = \Omega \times \mathcal{I}$ with $\mathbf{x} = (x_1, x_2) \in \Omega$, $\Omega \subset \mathcal{R}^2$ and $x_3 \in \mathcal{I}$, $\mathcal{I} = [0, H] \subset \mathcal{R}$, being H the plate thickness. The separated representation of the displacement field $\mathbf{u} = (u_1, u_2, u_3)$ reads:

$$\mathbf{u}(x_1, x_2, x_3) = \begin{pmatrix} u_1(x_1, x_2, x_3) \\ u_2(x_1, x_2, x_3) \\ u_3(x_1, x_2, x_3) \end{pmatrix} \approx \sum_{i=1}^{N} \begin{pmatrix} P_1^i(x_1, x_2) \cdot T_1^i(x_3) \\ P_2^i(x_1, x_2) \cdot T_2^i(x_3) \\ P_3^i(x_1, x_2) \cdot T_3^i(x_3) \end{pmatrix}, \quad (123)$$

where P_k^i, $k = 1, 2, 3$, are functions of the in-plane coordinates $\mathbf{x} = (x_1, x_2)$ whereas T_k^i, $k = 1, 2, 3$, are functions involving the thickness coordinate x_3.

Expression (123) can be written in a more compact form by using the Hadamard product:

$$\mathbf{u}(\mathbf{x}, x_3) \approx \sum_{i=1}^{N} \mathbf{P}^i(\mathbf{x}) \circ \mathbf{T}^i(x_3), \quad (124)$$

where vectors $\mathbf{P}^i$ and $\mathbf{T}^i$ contains functions P_k^i and T_k^i respectively.

Because neither the number of terms in the separated representation of the displacement field nor the dependence on x_3 of functions T_k^i are assumed *a priori*, the approximation is flexible enough for representing the fully 3D solution, being obviously more general than theories assuming particular *a priori* evolutions in the thickness direction x_3.

Let us consider a linear elasticity problem on a plate domain $\Xi = \Omega \times \mathcal{I}$. The weak formulation reads:

$$\int_{\Xi} \epsilon(\mathbf{u}^*)^T \cdot \mathbf{K} \cdot \epsilon(\mathbf{u}) \, d\mathbf{x} = \int_{\Xi} \mathbf{u}^* \cdot \mathbf{f}_d \, d\mathbf{x} + \int_{\Gamma_N} \mathbf{u}^* \cdot \mathbf{F}_d \, d\mathbf{x}, \quad (125)$$

with $\mathbf{K}$ the generalized 6×6 Hooke tensor, $\mathbf{f}_d$ represents the volumetric body forces while $\mathbf{F}_d$ represents the traction applied on the boundary Γ_N. In what follows we assume that $\mathbf{K}$, $\mathbf{f}_d$ and $\mathbf{F}_d$ accepts an in-plane-out-of-plane separated representation (we come back to this issue later).

The separated representation constructor proceeds by computing a term of the sum at each iteration. Assuming that the first $n-1$ modes (terms of the finite sum) of the solution were already computed, $\mathbf{u}^{n-1}(\mathbf{x}, x_3)$ with

$n \geq 1$, the solution enrichment reads:

$$\mathbf{u}^n(\mathbf{x}, x_3) = \mathbf{u}^{n-1}(\mathbf{x}, x_3) + \mathbf{P}^n(\mathbf{x}) \circ \mathbf{T}^n(x_3), \tag{126}$$

where both vectors $\mathbf{P}^n$ and $\mathbf{T}^n$ containing functions P_i^n and T_i^n ($i = 1, 2, 3$) depending on $\mathbf{x}$ and x_3 respectively, are unknown at the present iteration, resulting in a non-linear problem.

We proceed by considering the simplest linearization strategy, an alternating directions fixed point algorithm, that proceeds by calculating $\mathbf{P}^n$ by assuming $\mathbf{T}^n$ known, and then by updating $\mathbf{T}^n$ from the just calculated $\mathbf{P}^n$. The iteration procedure continues until convergence, that is, until reaching the fixed point.

When $\mathbf{T}^n$ is assumed known, we consider the test function $\mathbf{u}^\star$ given by $\mathbf{P}^\star \circ \mathbf{T}^n$. By introducing the trial and test functions into the weak form and then integrating in $\mathcal{I}$ because all the functions depending on the thickness coordinate are known, we obtain a 2D weak formulation defined in Ω whose discretization (by using a standard discretization strategy, e.g. finite elements) allows computing $\mathbf{P}^n$.

Analogously, when $\mathbf{P}^n$ is assumed known, the test function $\mathbf{u}^\star$ is given by $\mathbf{P}^n \circ \mathbf{T}^\star$. By introducing the trial and test functions into the weak form and then integrating in Ω because all the functions depending on the in-plane coordinates $\mathbf{x}$ are at present known, we obtain a 1D weak formulation defined in $\mathcal{I}$ whose discretization (using any technique for solving standard ODE equations) allows computing $\mathbf{T}^n$.

As discussed in Bognet et al. (2012) this separated representation allows computing 3D solutions while keeping a computational complexity characteristic of 2D solution procedures. If we consider a hexahedral domain discretized using a regular structured grid with N_1, N_2 and N_3 nodes in the x_1, x_2 and x_3 directions respectively, usual mesh-based discretization strategies imply a challenging issue because the number of nodes involved in the model scales with $N_1 \cdot N_2 \cdot N_3$, however, by using the separated representation and assuming that the solution involves N modes, one must solve about N 2D problems related to the functions involving the in-plane coordinates $\mathbf{x}$ and the same number of 1D problems related to the functions involving the thickness coordinate x_3. The computing time related to the solution of the one-dimensional problems can be neglected with respect to the one required for solving the two-dimensional ones. Thus, the resulting complexity scales as $N \cdot N_1 \cdot N_2$. By comparing both complexities we can notice that as soon as $N_3 \gg N$ the use of separated representations leads to impressive computing time savings, making possible the solution of models never until now solved, even the parametric ones, and using light computing platforms as illustrated in Bognet et al. (2012).

Elastic problem in shell domains

The shell domain Ω^S, assumed with constant thickness, can be described from a reference surface $\mathbf{X}$, that in what follows will be identified to the shell middle surface but that in the general case could be any other one, parametrized by the coordinates ξ, η, that is $\mathbf{X}(\xi, \eta)$, where:

$$\mathbf{X}(\xi,\eta) = \begin{pmatrix} X_1(\xi,\eta) \\ X_2(\xi,\eta) \\ X_3(\xi,\eta) \end{pmatrix}. \tag{127}$$

Being $\mathbf{n}$ the unit vector normal to the middle surface, the shell domain Ω^S can be parametrized from:

$$\mathbf{x}(\xi,\eta,\zeta) = \mathbf{X}(\xi,\eta) + \zeta \cdot \mathbf{n}. \tag{128}$$

The geometrical transformation $(\xi,\eta,\zeta) \rightarrow (x_1, x_2, x_3)$ involves

$$\tilde{\mathbf{F}} = \left[\frac{\partial \mathbf{x}}{\partial \xi} \ \frac{\partial \mathbf{x}}{\partial \eta} \ \mathbf{n} \right]. \tag{129}$$

With the weak form of the elastic problem defined in the reference domain $\Xi = \Omega \times \mathcal{I}$, with $(\xi,\eta) \in \Omega$ and $\zeta \in \mathcal{I}$, the situation is quite similar to the one encountered in the analysis of elastic problems in plate geometries addressed in Bognet et al. (2012).

In Bognet et al. (2014) we considered the in-plane-out-of-plane separated representation of the displacement field, similar to (101) but now involving the coordinates (ξ,η,ζ)

$$\mathbf{u}(\xi,\eta,\zeta) = \begin{pmatrix} u_1(\xi,\eta,\zeta) \\ u_2(\xi,\eta,\zeta) \\ u_3(\xi,\eta,\zeta) \end{pmatrix} \approx \sum_{i=1}^{N} \begin{pmatrix} P_1^i(\xi,\eta) \cdot T_1^i(\zeta) \\ P_2^i(\xi,\eta) \cdot T_2^i(\zeta) \\ P_3^i(\xi,\eta) \cdot T_3^i(\zeta) \end{pmatrix}, \tag{130}$$

or in a more compact form

$$\mathbf{u}(\xi,\eta,\zeta) \approx \sum_{i=1}^{N} \mathbf{P}^i(\xi,\eta) \circ \mathbf{T}^i(\zeta). \tag{131}$$

Darcy's flow model

We now illustrate the application of separated representations to the modeling of resin transfer molding Resin Transfer Molding – RTM – processes deeply addressed in Chinesta, Ammar et al. (2011). We consider the flow within a porous medium in a plate domain $\Xi = \Omega \times \mathcal{I}$ with $\Omega \subset \mathcal{R}^2$ and

$\mathcal{I} = [0, H] \subset \mathcal{R}$. The governing equation is obtained by combining Darcy's law, which relates the fluid velocity to the pressure gradient

$$\mathbf{v} = -\mathbf{K} \cdot \nabla p, \tag{132}$$

and the incompressibility constraint,

$$\nabla \cdot \mathbf{v} = 0. \tag{133}$$

Introduction of Eq. (132) into Eq. (133) yields a single equation for the pressure field

$$\nabla \cdot (\mathbf{K} \cdot \nabla p) = 0. \tag{134}$$

Remark: The heat equation being formally similar to Eq. (134) the considerations that follow also apply for the solution of thermal models in plate geometries. □

Again we assume the permeability separability

$$\mathbf{K}(x, y, z) = \sum_{i=1}^{P} \mathbf{K}_i(\mathbf{x}) \cdot \xi_i(z), \tag{135}$$

where $\mathbf{x}$ denotes the in-plane coordinates, i.e. $\mathbf{x} = (x, y) \in \Omega$.

The weak form of Eq. (134) reads:

$$\int_{\Xi} \nabla p^* \cdot (\mathbf{K} \cdot \nabla p) \ d\Xi = 0, \tag{136}$$

for all test functions p^* selected in an appropriate functional space. Dirichlet boundary conditions are imposed for the pressure at the inlet and outlet of the flow domain, while zero flux (i.e. no flow) is imposed elsewhere as a natural boundary condition. We seek an approximate solution $p(x, y, z)$ in the separated form:

$$p(\mathbf{x}, z) \approx \sum_{j=1}^{N} X_j(\mathbf{x}) \cdot Z_j(z). \tag{137}$$

The PGD algorithm then proceeds as follows. Assume that the first n functional products have been computed, i.e.

$$p^n(\mathbf{x}, z) = \sum_{j=1}^{n} X_j(\mathbf{x}) \cdot Z_j(z), \tag{138}$$

is a known quantity. We must now perform an enrichment step to obtain

$$p^{n+1}(\mathbf{x}, z) = p^n(\mathbf{x}, z) + R(\mathbf{x}) \cdot S(z). \tag{139}$$

The test function involved in the weak form is given by

$$p^*(\mathbf{x}, z) = R^*(\mathbf{x}) \cdot S(z) + R(\mathbf{x}) \cdot S^*(z). \tag{140}$$

Introducing Eqs. (139) and (140) into Eq. (136), we obtain

$$\int_\Xi \left(\begin{pmatrix} \tilde{\nabla} R^* \cdot S \\ R^* \cdot \frac{dS}{dz} \end{pmatrix} + \begin{pmatrix} \tilde{\nabla} R \cdot S^* \\ R \cdot \frac{dS^*}{dz} \end{pmatrix} \right) \cdot \left(\mathbf{K} \cdot \begin{pmatrix} \tilde{\nabla} R \cdot S \\ R \cdot \frac{dS}{dz} \end{pmatrix} \right) \, d\Xi =$$

$$= - \int_\Xi \left(\begin{pmatrix} \tilde{\nabla} R^* \cdot S \\ R^* \cdot \frac{dS}{dz} \end{pmatrix} + \begin{pmatrix} \tilde{\nabla} R \cdot S^* \\ R \cdot \frac{dS^*}{dz} \end{pmatrix} \right) \cdot \mathbf{Q}^n \, d\Xi, \tag{141}$$

where $\tilde{\nabla}$ denotes the plane component of the gradient operator, i.e. $\tilde{\nabla}^T = \left(\frac{\partial}{\partial x}, \frac{\partial}{\partial y} \right)$ and $\mathbf{Q}^n$ is a flux term known at step n:

$$\mathbf{Q}^n = \mathbf{K} \cdot \sum_{j=1}^{n} \begin{pmatrix} \tilde{\nabla} X_j(\mathbf{x}) \cdot Z_j(z) \\ X_j(\mathbf{x}) \cdot \frac{dZ_j(z)}{dz} \end{pmatrix}. \tag{142}$$

As discussed previously, each enrichment step of the PGD algorithm is a non-linear problem which must be performed by means of a suitable iterative process. The simplest one proceeds assuming $R(\mathbf{x})$ known to obtain $S(z)$, and then updating $R(\mathbf{x})$. The process continues until reaching convergence. The converged solutions provide the next functional product of the PGD: $R(\mathbf{x}) \rightarrow X_{n+1}(\mathbf{x})$ and $S(z) \rightarrow Z_{n+1}(z)$.

Stokes flow model

The Stokes model is defined in $\Xi = \Omega \times \mathcal{I}$ and reads, for an incompressible fluid:

$$\begin{cases} \nabla p = \nabla \cdot (\eta \cdot \nabla \mathbf{v}) \\ \nabla \cdot \mathbf{v} = 0 \end{cases}. \tag{143}$$

To circumvent the issue related to stable mixed formulations (LBB conditions) within the separated representation we consider a penalty formulation that modifies the mass balance by introducing a penalty coefficient λ small enough

$$\nabla \cdot \mathbf{v} + \lambda \cdot p = 0, \tag{144}$$

or more explicitly

$$p = -\frac{1}{\lambda} \left(\frac{\partial u}{\partial x} + \frac{\partial v}{\partial y} + \frac{\partial w}{\partial z} \right) = -\frac{\nabla \cdot \mathbf{v}}{\lambda}. \tag{145}$$

By replacing it into the balance of momentum (first equation in (143)) we obtain

$$\nabla(\nabla\cdot\mathbf{v})+\xi\Delta\mathbf{v}=0, \tag{146}$$

with $\xi=\eta\cdot\lambda$.

In the present case it suffices considering the separated representation of the velocity field according to Ghnatios et al. (2014):

$$\mathbf{v}=\begin{pmatrix}u\\v\\w\end{pmatrix}\approx\begin{pmatrix}\sum\limits_{i=1}^{i=N}X_i^u(x,y)\cdot Z_i^u(z)\\\sum\limits_{i=1}^{i=N}X_i^v(x,y)\cdot Z_i^v(z)\\\sum\limits_{i=1}^{i=N}X_i^w(x,y)\cdot Z_i^w(z)\end{pmatrix}, \tag{147}$$

that leads to a separated representation of the strain rate, that introduced into the Stokes problem weak form allows the calculation of functions $X_i(x,y)$ by solving the corresponding 2D problems and functions $Z_i(z)$ by solving the associated 1D problems. Because of the one-dimensional character of problems defined in the laminate thickness we can use extremely fine descriptions along the thickness direction without a significant impact on the computational efficiency.

4.2 Laminates

In the case of laminates, several plies with different thermomechanical properties (eventually anisotropic) are found through the domain thickness. In the case of addressing the flow in a porous preform we could suppose P different anisotropic plies of thickness h, each one characterized by a permeability tensor $\mathbf{K}_i(x,y)$ that is assumed constant through the ply thickness. Then, we define a characteristic function

$$\xi_i(z)=\begin{cases}1 & z_i\leq z\leq z_{i+1}\\0 & otherwise\end{cases}, \tag{148}$$

where $z_i=(i-1)\cdot h$ is the location of ply i in the plate thickness. The laminate's permeability is thus given in separated form

$$\mathbf{K}(x,y,z)=\sum_{i=1}^{P}\mathbf{K}_i(\mathbf{x})\cdot\xi_i(z), \tag{149}$$

where $\mathbf{x}$ denotes the in-plane coordinates, i.e. $\mathbf{x}=(x,y)\in\Omega$.

Brinkman model

In composites manufacturing processes resin located among the fibers in the reinforcement layers also flows. A usual approach for evaluating the resin flow in such circumstances consists of solving the associated Darcy's model. It is well known that Darcy-Stokes coupling at the interlayers generates numerical instabilities because the localized boundary layers whose accurate description requires very rich representations (very fine meshes along the laminate thickness).

In Ghnatios et al. (2014) we proposed the use of the Brinkman model that allows representing in an unified manner both the Darcy and the Stokes behaviors. In order to avoid numerical inaccuracies we use a very fine representation along the thickness direction and for circumventing the exponential increase in the number of degrees of freedom that such a fine representation would imply when extended to the whole laminate domain, we consider again the in-plane-out-of-plane separated representation previously introduced.

The Brinkman model is defined by:

$$\nabla p = \mu \cdot \mathbf{K}^{-1} \cdot \mathbf{v} + \eta \cdot \Delta \mathbf{v}, \tag{150}$$

where μ is the dynamic viscosity, $\mathbf{K}$ the layer permeability tensor and η the dynamic effective viscosity.

In the zones where Stokes model applies (resin layers) we assign a very large isotropic permeability $K = 1$ (units in the metric system) whereas in the ones occupied by the reinforcement, the permeability is assumed anisotropic, being several orders of magnitude lower, typically 10^{-8}. Thus the Darcy's component in Eq. (150) does not perturb the Stokes flow in the resin layers, and it becomes dominant in the reinforcement layers.

On the approximation continuity

All the just addressed models imply second order derivatives in the space coordinates, and then, after integrating by parts to recover their associated weak forms, only continuous approximations are required for both the trial and the test functions. Thus, in our numerical applications we considered standard piecewise linear functions for approximating the 2D fields defined in Ω and the 1D defined in $\mathcal{I}$. Obviously higher order approximations are possible with the only constraint of ensuring continuity.

Electromagnetic models derived from Maxwell's equations usually involves vector potentials that appears in the weak form affected by the *curl* operator. Let's $\mathbf{A} = (A_x, A_y, A_z)$ such a potential. We have

$$(\nabla \times \mathbf{A})^T = \left(\frac{\partial A_z}{\partial y} - \frac{\partial A_y}{\partial z}, \frac{\partial A_x}{\partial z} - \frac{\partial A_z}{\partial x}, \frac{\partial A_y}{\partial x} - \frac{\partial A_x}{\partial y} \right), \tag{151}$$

that proves that A_x must be continuous with respect to the coordinates y and z but should be preferably discontinuous in the x coordinate in order to ensure the transfer conditions from one medium to other. The same reasoning applies for the other components of the vector potential.

If for the sake of clarity we consider in what follows the 2D case defined by the coordinates (x, z) we could approximate A_x and A_z from

$$\begin{cases} A_x \approx \sum_{i=1}^{i=N} X_i^x(x) \cdot Z_i^x(z) \\ A_z \approx \sum_{i=1}^{i=N} X_i^z(x) \cdot Z_i^z(z) \end{cases}, \tag{152}$$

where X_i^x and Z_i^x are approximated using piecewise constant (discontinuous) and linear (continuous) functions of the x and z coordinates respectively. On the other hand X_i^z and Z_i^z are approximated using piecewise linear (continuous) and constant (discontinuous) functions of the x and z coordinates respectively. This simple choice ensure the continuity requirement just specified, and constitutes the simplest generalization of the so-called Nedelec's elements usually considered in electromagnetic numerical simulations.

5 Applications

5.1 Multidimensional models

As discussed previously, some models are inherently defined in a configurational or conformation space that is high-dimensional in nature. Its discretization leads to the well-known curse of dimensionality if traditional, mesh-based techniques are applied.

Separated representations were applied for solving the multidimensional Fokker-Planck equation describing complex fluids in the kinetic theory framework. In Ammar, Mokdad et al. (2006), the authors addressed the solution of the linear Fokker-Planck equation describing multi-bead-spring molecular models in the steady state and in homogeneous flows. In that case the distribution function describing the molecular conformation only depends on the molecule configurational coordinates $\mathbf{q}_1, \cdots \mathbf{q}_D$ (see Fig. 6). The molecular conformation distribution function $\Psi(\mathbf{q}_1, \cdots, \mathbf{q}_D)$ results from the solution of the associated balance Fokker-Planck equation:

$$\frac{\partial \Psi}{\partial t} = - \sum_{i=1}^{i=D} \left(\dot{\mathbf{q}}_i \cdot \Psi \right), \tag{153}$$

where the conformational velocities $\dot{\mathbf{q}}_i$ depend on the molecule conformation and the molecule-fluid interactions. The solution was searched in the separated form:

$$\Psi(\mathbf{q}_1, \cdots, \mathbf{q}_D) \approx \sum_{i=1}^{i=N} F_i^1(\mathbf{q}_1) \times \cdots \times F_i^D(\mathbf{q}_D). \tag{154}$$

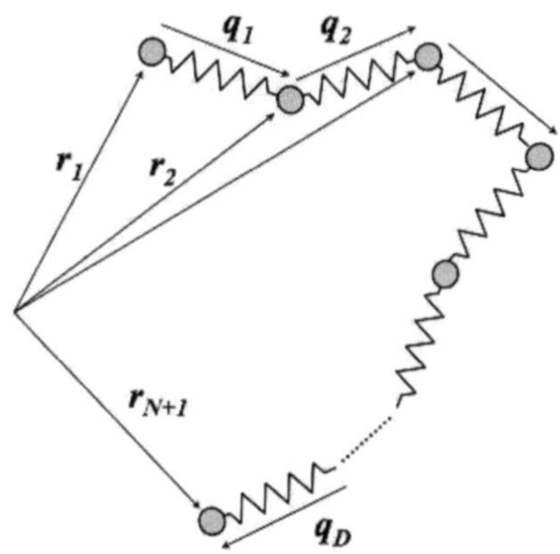

Figure 6. Molecular conformation

The solution procedure was extended to non-linear kinetic theory descriptions of more complex molecular descriptions in Mokdad et al. (2007). The transient solution was addressed in Ammar et al. (2007) in which the time was added as an extra-coordinate. Transient solutions making use of reduced bases in the context of an adaptive proper orthogonal decomposition (Ryckelynck et al. (2006)) were considered in the case of low dimensional configuration spaces: the FENE model was addressed in Ammar, Ryckelynck et al. (2006) and liquid crystalline polymers in Ammar et al. (2009). Complex fluid models were coupled with complex flows in Pruliere et al. (2009) and in Mokdad et al. (2010) opening very encouraging perspectives and claiming the necessity of defining efficient stabilizations. Finally, in Chinesta et al. (2007) the PGD was applied for solving the stochastic equation within the Brownian Configuration Field framework. The interested reader can refer to Chinesta et al. (2011) and the references therein for an exhaustive overview of PGD in computational rheology.

Multidimensional models encountered in the finer descriptions of matter (ranging from quantum chemistry to statistical mechanics descriptions) were revisited in Ammar et al. (2008). The multidimensional chemical master equation was efficiently solved in Ammar, Cueto and Chinesta (2012). Langer's equation governing phase transitions was considered in Lamari et al. (2012).

5.2 Parametric models: a route to efficient optimization, inverse identification and real time simulation

Usual computational mechanics models could be enriched by introducing several extra-coordinates. Thus, adding some new coordinates to models initially non high-dimensional, could lead to new, never "before" explored insights in the physics as previously illustrated.

This kind of parametric modeling was addressed in Ammar, Normandin and Chinesta (2010), Pruliere, Chinesta and Ammar (2010) and Bognet et al. (2012) where material parameters were introduced as extra-coordinates. In Lamari et al. (2010) thermal conductivities, macroscopic temperature and its time evolution were introduced as extra-coordinates for performing linear and non-linear homogenization.

In Bognet (2012) the anisotropy directions of plies involved in a composite laminate were considered as extra-coordinates. By assuming a certain uncertainty in the real orientation of such plies, authors evaluated the envelope of the resulting distorted structures due to the thermomechanical coupling.

Moreover, as soon as the separated representation of the parametric solution has been computed off-line, its on-line use only needs to particularize such solutions for a desired set of parameters. Obviously, this task can be performed very fast, many times in real time, and by using light computing platforms, as smartphones or tablets.

The interested reader can refer to Chinesta et al. (2013) and the references therein for an exhaustive review on PGD-based parametric solutions.

5.3 Real-time simulation, DDDAS and more

It is easy to understand that after performing this type of calculations, in which parameters are considered advantageously as new coordinates of the model, a posteriori inverse identification or optimization can be easily handled. This new PGD framework allows us to perform this type of calculations very efficiently, because in fact all possible solutions have been previously computed in the form of a separated, high-dimensional solution so that they constitute a simple post-processing of this general solution.

Process optimization was considered in Ghnatios et al. (2011) for instance. Shape optimization was performed by considering all the geometrical parameters as extra-coordinates, leading to the model solution in any of the geometries generated by the parameters considered as extra-coordinates, the plies thickness in a laminate (Leygue and Verron (2010)). More complex scenarios were considered in Ammar, Huerta et al. (2012) where the Laplace equation was solved in a parametrized domain.

Inverse methods in the context of real-time simulations were addressed in Gonzalez et al. (2012) and were coupled with control strategies in Ghnatios et al. (2012) as a first step towards DDDAS (dynamic data-driven application systems). Moreover, because the general parametric solution was pre-computed off-line, it can be used on-line under real time constraints and using light computing platforms like smartphones (Bognet et al. (2012) and Ghnatios et al. (2012)) that constitute a first step towards augmented reality applications.

As mentioned before, surgical simulators must operate at frequencies higher than 500 Hz. The use of model reduction seems to be an appealing alternative for reaching such performances. However, techniques based on the use of the POD and PODI, even combined with an asymptotic numerical methods to avoid the computation of the tangent matrix, exhibit serious difficulties to fulfill such requirements as discussed in the Niroomandi's works. PGD constitutes the most appealing route for circumventing the computational challenges.

Bibliography

[1] A. Ammar, B. Mokdad, F. Chinesta, R. Keunings. A new family of solvers for some classes of multidimensional partial differential equations encountered in kinetic theory modeling of complex fluids. Journal of Non-Newtonian Fluid Mechanics, 139, 153-176, 2006.

[2] A. Ammar, D. Ryckelynck, F. Chinesta, R. Keunings. On the reduction of kinetic theory models related to finitely extensible dumbbells. Journal of Non-Newtonian Fluid Mechanics, 134, 136-147, 2006.

[3] A. Ammar, B. Mokdad, F. Chinesta, R. Keunings. A new family of solvers for some classes of multidimensional partial differential equations encountered in kinetic theory modeling of complex fluids. Part II: Transient simulation using space-time separated representation. Journal of Non-Newtonian Fluid Mechanics, 144, 98-121, 2007.

[4] A. Ammar, F. Chinesta, P. Joyot. The nanometric and micrometric

scales of the structure and mechanics of materials revisited: An introduction to the challenges of fully deterministic numerical descriptions. International Journal for Multiscale Computational Engineering, 6/3, 191-213, 2008.

[5] A. Ammar, E. Pruliere, F. Chinesta, M. Laso. Reduced numerical modeling of flows involving liquid-crystalline polymeres. Journal of Non-Newtonian Fluid Mechanics, 160, 140-156, 2009.

[6] A. Ammar, E. Pruliere, J. Ferec, F. Chinesta, E. Cueto. Coupling finite elements and reduced approximation bases. European Journal of Computational Mechanics, 18/5-6, 445-463, 2009.

[7] A. Ammar, M. Normandin, F. Daim, D. Gonzalez, E. Cueto, F. Chinesta. Non-incremental strategies based on separated representations: Applications in computational rheology. Communications in Mathematical Sciences, 8/3, 671-695, 2010.

[8] A. Ammar, F. Chinesta, P. Diez, A. Huerta. An error estimator for separated representations of highly multidimensional models. Computer Methods in Applied Mechanics and Engineering, 199 1872-1880, 2010.

[9] A. Ammar, M. Normandin, F. Chinesta. Solving parametric complex fluids models in rheometric flows. Journal of Non-Newtonian Fluid Mechanics, 165, 1588-1601, 2010.

[10] A. Ammar, F. Chinesta, E. Cueto. Coupling finite elements and Proper Generalized Decomposition. International Journal of Multiscale Computational Engineering, 9/1, 17-33, 2011.

[11] A. Ammar, F. Chinesta, E. Cueto, M. Doblare. Proper Generalized Decomposition of time-multiscale models. International Journal for Numerical Methods in Engineering, 90/5, 569-596, 2012.

[12] A. Ammar, E. Cueto, F. Chinesta. Reduction of the Chemical Master Equation for gene regulatory networks using Proper Generalized Decompositions. International Journal for Numerical Methods in Biomedical Engineering, 28/9, 960-973, 2012.

[13] A. Ammar, A. Huerta, F. Chinesta, E. Cueto, A. Leygue. Parametric solutions involving geometry: A step towards efficient shape optimization. Computer Methods in Applied Mechanics and Engineering, 268, 178-193, 2014.

[14] M Barrault, Y Maday, NC Nguyen, A.T Patera. An "empirical interpolation" method: application to efficient reduced-basis discretization of partial differential equations. Comptes Rendus Mathematique, 339/9, 667-672, 2004.

[15] N. Bellomo, Modeling complex living systems, Birkhauser, 2008.

[16] Ch. Bernoulli, Vademecum des Mechanikers. Cotta, Stuttgart, 1836.

[17] R.A. Bialecki, A.J. Kassab, A. Fic, Proper orthogonal decomposition and modal analysis for acceleration of transient FEM thermal analysis, Int. J. Numer. Meth. Engrg., 62, 774-797, 2005.

[18] B.B. Bird, C.F. Curtiss, R.C. Armstrong, O. Hassager, Dynamics of polymeric liquids, in: Kinetic Theory, vol. 2, John Wiley & Sons, 1987.

[19] B. Bognet, A. Leygue, F. Chinesta, A. Poitou, F. Bordeu. Advanced simulation of models defined in plate geometries: 3D solutions with 2D computational complexity. Computer Methods in Applied Mechanics and Engineering, 201, 1-12, 2012.

[20] B. Bognet, A. Leygue, F. Chinesta. Separated representations of 3D elastic solutions in shell geometries. Advanced Modelling and Simulation in Engineering Sciences, In press, 2014.

[21] T. Bui-Thanh, K. Willcox, O. Ghattas, B. Van Bloemen Waanders. Goal-oriented, model-constrained optimization for reduction of large-scale systems. Journal of Computational Physics, 224/2, 880-896, 2007.

[22] J. Burkardt, M. Gunzburger, H-Ch. Lee, POD and CVT-based reduced-order modeling of Navier-Stokes flows, Comput. Methods Appl. Mech. Engrg., 196; 337-355, 2006.

[23] E. Cancès, M. Defranceschi, W. Kutzelnigg, C. Le Bris, Y. Maday, Computational Quantum Chemistry: a primer, Handbook of Numerical Analysis, Elsevier, Vol. X, 3-270, 2003.

[24] S. Chaturantabut, D.C. Sorensen. Nonlinear model reduction via discrete empirical interpolation. SIAM J. Sci. Comput., 32, 2737-2764, 2010.

[25] F. Chinesta, A. Ammar, A. Falco, M. Laso. On the reduction of stochastic kinetic theory models of complex fluids. Modeling and Simulation in Materials Science and Engineering, 15, 639-652, 2007.

[26] F. Chinesta, A. Ammar, F. Lemarchand, P. Beauchene, F. Boust. Alleviating mesh constraints: Model reduction, parallel time integration and high resolution homogenization. Computer Methods in Applied Mechanics and Engineering, 197/5, 400-413, 2008.

[27] F. Chinesta, A. Ammar, E. Cueto. Proper Generalized Decomposition of multiscale models. International Journal for Numerical Methods in Engineering, 83/8-9, 1114-1132, 2010.

[28] F. Chinesta, A. Ammar, E. Cueto. Recent advances and new challenges in the use of the Proper Generalized Decomposition for solving multidimensional models. Archives of Computational Methods in Engineering, 17/4, 327-350, 2010.

[29] F. Chinesta, A. Ammar, A. Leygue, R. Keunings. An overview of the Proper Generalized Decomposition with applications in computational rheology. Journal of Non Newtonian Fluid Mechanics, 166, 578-592, 2011.

[30] F. Chinesta, P. Ladeveze, E. Cueto. A short review in model order reduction based on Proper Generalized Decomposition. Archives of Computational Methods in Engineering, 18, 395-404, 2011.

[31] F. Chinesta, A. Leygue, F. Bordeu, J.V. Aguado, E. Cueto, D. Gonzalez, I. Alfaro, A. Ammar, A. Huerta. Parametric PGD based computational vademecum for efficient design, optimization and control. Archives of Computational Methods in Engineering, 20/1, 31-59, 2013.

[32] F. Chinesta, R. Keunings, A. Leygue. The Proper Generalized Decomposition for Advanced Numerical Simulations. A primer. Springerbriefs, Springer, 2014.

[33] F. Chinesta, A. Leygue, B. Bognet, Ch. Ghnatios, F. Poulhaon, F. Bordeu, A. Barasinski, A. Poitou, S. Chatel, S. Maison-Le-Poec. First steps towards an advanced simulation of composites manufacturing by Automated Tape Placement. International Journal of Material Forming, In press, 2014.

[34] B. Cochelin, N. Damil, M. Potier-Ferry. The asymptotic numerical method: an efficient perturbation technique for nonlinear structural mechanics. Revue Europeenne des Elements Finis, 3, 281-297, 1994.

[35] F. Darema, Engineering/Scientific and Commercial applications: differences, similarities, and future evolution, Proceedings of the Second

Hellenic European Conference on Mathematics and Informatics. HERMIS, 1, 367-374, 1994.

[36] J. Donea, A. Huerta. Finite element methods for flow problems. J. Wiley and Sons, 2002.

[37] J.L Eftang, D.J Knezevic, A.T Patera, An hp Certified Reduced Basis Method for Parametrized Parabolic Partial Differential Equations. Mathematical and Computer Modelling of Dynamical Systems, 17/4, 395-422, 2011.

[38] Ch. Ghnatios, F. Chinesta, E. Cueto, A. Leygue, P. Breitkopf, P. Villon. Methodological approach to efficient modeling and optimization of thermal processes taking place in a die: Application to pultrusion. Composites Part A, 42, 1169-1178, 2011.

[39] Ch. Ghnatios, F. Masson, A. Huerta, E. Cueto, A. Leygue, F. Chinesta. Proper Generalized Decomposition based dynamic data-driven control of thermal processes. Computer Methods in Applied Mechanics and Engineering, 213, 29-41, 2012.

[40] Ch. Ghnatios, F. Chinesta, Ch. Binetruy. The squeeze flow of composite laminates. International Journal of Material Forming, In press, 2014.

[41] M. Girault, E. Videcoq, D. Petit. Estimation of time-varying heat sources through inversion of a low order model built with the Modal Identification Method from in-situ temperature measurements. International Journal of Heat and Mass Transfer, 53, 206-219, 2010.

[42] D. Gonzalez, A. Ammar, F. Chinesta, E. Cueto. Recent advances in the use of separated representations. International Journal for Numerical Methods in Engineering, 81/5, 637-659, 2010.

[43] D. Gonzalez, F. Masson, F. Poulhaon, A. Leygue, E. Cueto, F. Chinesta. Proper Generalized Decomposition based dynamic data-driven inverse identification. Mathematics and Computers in Simulation, 82/9, 1677-1695, 2012.

[44] D. Gonzalez, E. Cueto, F. Chinesta, P. Diez, A. Huerta. SUPG-based stabilization of Proper Generalized Decompositions for high-dimensional advection-diffusion equations. International Journal for Numerical Methods in Engineering, 94/13, 1216-1232, 2013.

[45] M.D. Gunzburger, J.S. Peterson, J.N. Shadid, Reduced-order modeling of time-dependent PDEs with multiple parameters in the boundary data. Comput. Methods Appl. Mech. Engrg., 196, 1030-1047, 2007.

[46] T.J.R. Hughes, Multiscale phenomena: Green's functions, the Dirichlet-to-Neumann formulation, subgrid scale models, bubbles and the origins of stabilized methods, Computer Methods in Applied Mechanics and Engineering, 127/1-4, 387-401, 1995.

[47] T.J R. Hughes, G.R. Feijóo, L. Mazzei, J-B. Quincy. The variational multiscale method - a paradigm for computational mechanics. Comput. Methods Appl. Mech. Eng., 166/1-2, 3-24, 1998.

[48] P. Ladevèze, The large time increment method for the analyze of structures with nonlinear constitutive relation described by internal variables, Comptes Rendus Acad?mie des Sciences Paris, 309, 1095-1099, 1989.

[49] P. Ladeveze, Nonlinear computational structural mechanics, Springer, NY, 1999.

[50] P. Ladevèze, A. Nouy, A multiscale computational method with time and space homogenization, Comptes Rendus Mécanique, 330/10, 683-689, 2002.

[51] P. Ladevèze, A. Nouy, O. Loiseau, A multiscale computational approach for contact problems, Computer Methods In Applied Mechanics and Engineering, 191/43, 4869-4891, 2002.

[52] P. Ladevèze, A. Nouy, On a multiscale computational strategy with time and space homogenization for structural mechanics, Computer Methods In Applied Mechanics and Engineering, 192/28-30, 3061-3087, 2003.

[53] P. Ladevèze, D. Néron, P. Gosselet, On a mixed and multiscale domain decomposition method, Computer Methods in Applied Mechanics and Engineering, 96, 1526-1540, 2007.

[54] P. Ladevèze, J.-C. Passieux, D. Néron, The latin multiscale computational method and the proper generalized decomposition, Computer Methods In Applied Mechanics and Engineering, 199/21-22, 1287-1296, 2010.

[55] P. Ladevèze, L. Chamoin, On the verification of model reduction methods based on the proper generalized decomposition, Computer Methods In Applied Mechanics and Engineering, 200, 2032-2047, 2011.

[56] H. Lamari, A. Ammar, P. Cartraud, G. Legrain, F. Jacquemin, F. Chinesta. Routes for Efficient Computational Homogenization of

Non-Linear Materials Using the Proper Generalized Decomposition. Archives of Computational Methods in Engineering, 17/4, 373-391, 2010.

[57] H. Lamari, A. Ammar, A. Leygue, F. Chinesta. On the solution of the multidimensional Langer's equation by using the Proper Generalized Decomposition Method for modeling phase transitions. Modelling and Simulation in Materials Science and Engineering, 20, 015007, 2012.

[58] A. Leygue, E. Verron. A first step towards the use of Proper General Decomposition method for structural optimization. Archives of Computational Methods in Engineering, 17/4, 465-472, 2010.

[59] A. Leygue, F. Chinesta, M. Beringhier, T.L. Nguyen, J.C. Grandidier, F. Pasavento, B. Schrefler. Towards a framework for non-linear thermal models in shell domains. International Journal of Numerical Methods for Heat and Fluid Flow, 23/1, 55-73, 2013.

[60] Y. Maday, E.M. Ronquist. A reduced-basis element method. C. R. Acad. Sci. Paris, Ser. I, vol. 335, 195-200, 2002.

[61] Y. Maday, A.T. Patera, G. Turinici. A priori convergence theory for reduced-basis approximations of single-parametric elliptic partial differential equations. Journal of Scientific Computing, 17/1-4, 437-446, 2002.

[62] Y. Maday, E.M. Ronquist, The reduced basis element method: application to a thermal fin problem, SIAM J. Sci. Comput., 26/1, 240-258, 2004.

[63] B. Mokdad, E. Pruliere, A. Ammar, F. Chinesta. On the simulation of kinetic theory models of complex fluids using the Fokker-Planck approach. Applied Rheology,17/2, 26494, 1-14, 2007.

[64] B. Mokdad, A. Ammar, M. Normandin, F. Chinesta, J.R. Clermont. A fully deterministic micro-macro simulation of complex flows involving reversible network fluid models. Mathematics and Computer in Simulation, 80, 1936-1961, 2010.

[65] D. Néron, P. Ladevèze, Proper generalized decomposition for multiscale and multiphysics problems, Archives of Computational Methods In Engineering, 17/4, 351-372, 2010.

[66] S. Niroomandi, I. Alfaro, E. Cueto, F. Chinesta. Real-time deformable models of non-linear tissues by model reduction techniques. Computer Methods and Programs in Biomedicine, 91, 223-231, 2008.

[67] S. Niroomandi, I. Alfaro, E. Cueto, F. Chinesta. Model order reduction for hyperelastic materials. International Journal for Numerical Methods in Engineering, 81/9, 1180-1206, 2010.

[68] S. Niroomandi, I. Alfaro, E. Cueto, F. Chinesta. Accounting for large deformations in real-time simulations of soft tissues based on reduced order models. Computer Methods and Programs in Biomedicine, 105, 1-12, 2012.

[69] S. Niroomandi, I. Alfaro, D. Gonzalez, E. Cueto, F. Chinesta. Real time simulation of surgery by reduced order modelling and X-FEM techniques. International Journal for Numerical Methods in Biomedical Engineering, 28/5, 574-588, 2012.

[70] A. Nouy. Proper Generalized Decompositions and separated representations for the numerical solution of high dimensional stochastic problems. Archives of Computational Methods in Engineering - State of the Art Reviews, 17, 403-434, 2010.

[71] J.T. Oden, T. Belytschko, J. Fish, T.J.R. Hughes, C. Johnson, D. Keyes, A. Laub, L. Petzold, D. Srolovitz, S.Yip, Simulation-based Engineering Science: Revolutionizing Engineering Science through simulation. NSF Blue Ribbon Panel on SBES, 2006.

[72] H.M. Park, D.H. Cho, The use of the Karhunen-Loeve decomposition for the modelling of distributed parameter systems, Chem. Engineer. Science, 51, 81-98, 1996.

[73] J.-C. Passieux, P. Ladevèze, D.Néron, A scalable time-space multiscale domain decomposition method: Adaptive time scale separation, Computational Mechanics, 46/4, 621-633, 2010.

[74] E. Pruliere, A. Ammar, N. El Kissi, F. Chinesta. Recirculating flows involving short fiber suspensions: Numerical difficulties and efficient advanced micro-macro solvers. Archives of Computational Methods in Engineering, State of the Art Reviews, 16, 1-30, 2009.

[75] E. Pruliere, J. Ferec, F. Chinesta, A. Ammar. An efficient reduced simulation of residual stresses in composites forming processes. International Journal of Material Forming, 3/2, 1339-1350, 2010.

[76] E. Pruliere, F. Chinesta, A. Ammar. On the deterministic solution of multidimensional parametric models by using the Proper Generalized Decomposition. Mathematics and Computers in Simulation, 81, 791-810, 2010.

[77] E. Pruliere, F. Chinesta. A. Ammar, A. Leygue, A. Poitou. On the solution of the heat equation in very thin tapes. International Journals of Thermal Sciences, 65, 148-157, 2013.

[78] G. Rozza, D.B.P. Huynh, A.T. Patera. Reduced basis approximation and a posteriori error estimation for affinely parametrized elliptic coercive partial differential equations – application to transport and continuum mechanics. Archives of Computational Methods in Engineering, 15/3, 229-275, 2008.

[79] D. Ryckelynck, L. Hermanns, F. Chinesta, E. Alarcon. An efficient a priori model reduction for boundary element models. Engineering Analysis with Boundary Elements, 29, 796-801, 2005.

[80] D. Ryckelynck, A priori hyper-reduction method: an adaptive approach, Journal of Computational Physics, 202, 346-366, 2005.

[81] D. Ryckelynck, F. Chinesta, E. Cueto, A. Ammar. On the a priori model reduction: Overview and recent developments. Archives of Computational Methods in Engineering, State of the Art Reviews, 13/1, 91-128, 2006.

[82] D. Ryckelynck. Hyper-reduction of mechanical models involving internal variables. International Journal for Numerical Methods in Engineering, 77/1, 75-89, 2008.

[83] F. Schmidt, N. Pirc, M. Mongeau, F. Chinesta. Efficient mould cooling optimization by using model reduction. International Journal of Material Forming, 4/1, 71-82, 2011.

[84] K. Veroy, A. Patera. Certified real-time solution of the parametrized steady incompressible Navier-Stokes equations: Rigorous reduced-basis a posteriori error bounds. Int. J. Numer. Meth. Fluids, 47, 773-788, 2005.

[85] E. Videcoq, O. Quemener, M. Lazard, A. Neveu. Heat source identification and on-line temperature control by a Branch Eigenmodes Reduced Model. International Journal of Heat and Mass Transfer, 51 4743-4752, 2008.

PGD in linear and nonlinear Computational Solid Mechanics

Pierre Ladevèze

* LMT-Cachan (ENS Cachan/CNRS/UPMC/PRES UniverSud Paris)
61 avenue du Président Wilson, 94235 Cachan Cedex, France
ladeveze@lmt.ens-cachan.fr

Abstract Mechanics continues to supply numerous science and engineering problems which remain inaccessible to standard FE codes. Not all these problems are exotic, and many are indeed practical problems. A significant number of these engineering challenges are related to the today's growing interest in physics-based material models described on a scale smaller than that of the macroscopic structure, with applications such as structural design for which quasi real time simulation is mandatory. Design parameters and lacks of knowledge (variability, uncertainties) involving multiple parameters make these problems even more difficult.

This chapter addresses our answer to these computational challenges which is based on the Proper Generalized Decomposition (PGD) method that we have introduced in 1985 and developed until now. The two papers (Néron and Ladevèze, 2010; Chamoin et al., 2012) and the book (Ladevèze, 1996, 1999) are at the center of this chapter. Additions concern the main technical points which are detailed here for the first time.

Keywords PGD, reduced model, nonlinear, multiscale, verification, LATIN method, POD

Contents

F. Chinesta, P. Ladevèze (Eds.), *Separated Representations and PGD-Based Model Reduction Fundamentals and Applications,* CISM International Centre for Mechanical Sciences
DOI 10.1007/ 978-3-7091-1794-1_3

1 Introduction

1.1 Our main motivation

Our work is part of the "composites" revolution which the aeronautical industry has engaged in, particularly in Europe. The share of composites in civilian planes is growing from 20% to 50%, and vital elements such as the keel beam of the A380 or the fuselage and the wings of the A350 are made of carbon-fiber laminated composites. This escalation, along with the ever more stringent constraints in terms of energy consumption, safety and protection of the environment, make the current approach to analysis and sizing prohibitively expensive in cost and duration (and, therefore, inadequate) because it is based almost exclusively on testing. Today, industrializing a new matrix grade requires carrying out the whole series of tests with their numerous stacking sequences all over again, which inhibits innovation.

This is leading the aeronautical industry to a reversal of the situation through "Virtual Testing", which consists in replacing, whenever possible, the numerous experimental tests used today by virtual tests. Then experimental charts are replaced by virtual ones. This will lead to a significant decrease in the cost and duration of the design and sizing stage.

Advanced damage material models for composites exist today. They are able to well reproduce the damage physic until the final fracture but they lead to very-large-scale computational nonlinear problems. That is a big problem because engineering design needs quasi real time simulation. Our answer to this scientific challenge is to be able to derive offline PGD-Virtual Charts which are in fact reduced models related to several outputs of interest.

1.2 The PGD definition

The main approach we are developing in order to derive Virtual Charts is the Proper Generalized Decomposition (PGD). This could be seen as an

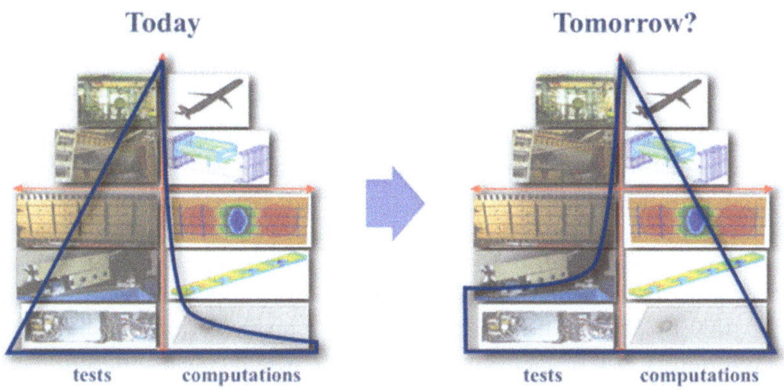

Figure 1. Future of the test/computation pyramid

extension of POD, which we introduced in (Ladevèze, 1985a,b) under the name "Radial Time-Space Approximation". Then, we have to solve offline very-large-scale nonlinear problems using all the available high performance tools (multiscale calculation strategies, parallel computing ...). Generally, PGD allows also to diminish drastically the computation and storage cost.

The main point with the PGD consists in calculating shape functions and the solution itself simultaneously using an iterative procedure. A priori, these shape functions are arbitrary and must only satisfy a variable separation assumption or something equivalent. That is a departure from other approximation methods for which shape functions are a priori given or partially given.

1.3 History of the PGD at LMT-Cachan

PGD fundamentals. The PGD/LATIN method was introduced for viscoplastic materials whose constitutive relations are described using a functional approach in (Ladevèze, 1985a,b). PGD is a global iterative computational approach over the time-space domain which lends itself to the separation of time and space variables. This is a departure from the incremental or step-by-step methods used traditionally. Indeed, PGD is not based on the concept of an increment; instead, it uses an iterative approach which, at each iteration, produces an approximation of the displacements, strains and stresses everywhere in the structure over the entire time interval of interest (which, a priori, is unlimited). Originally, PGD was called radial loading approximation, which, to us, meant a "mechanics" approximation in solid mechanics. In 2010, together with F. Chinesta, we changed the name because PGD can be viewed as an extension of the classical POD approach. All the presentations were done in the continuum mechanics framework, except for (Ladevèze and Perego, 2000), which uses the discretized framework. The extension of PGD to modern material descriptions involving internal variables, still for viscoplastic materials, was proposed in (Ladevèze, 1989, 1991). A number of mathematical properties regarding convergence and error indicators were proved in the book (Ladevèze, 1996, 1999). Over the years, several variations of the PGD computation technique have been derived in order to improve its robustness. A first important step was made around 1990 by Bussy and Cognard with the introduction of what is called the "preliminary stage". This is a POD stage which consists in updating the time functions , the shape functions of the space variable being kept. The corresponding algorithms are what mathematicians today call "greedy" algorithms. Only recently has a truly robust technique been found (Ladevèze et al., 2010). The extension of PGD to problems involving large displacements and large strains, which is not straightforward, was presented in (Ladevèze, 1991, 1996). The difficulties encountered, especially on the computational level, arise from the fact that the constitutive relations and the equilibrium equations are formulated in very different "worlds". The new formulation we proposed for PGD is a purely material (Lagrangian) formulation which unifies the different variables and equations. Moreover, it is formally similar to the small-displacement formulation. Another important issue is the treatment of nonlinear problems with parameters. A first approach is to use a remarkable property of the LATIN method which is that the iterative process can start from any time-space solution (Boucard and Ladevèze, 1999). Another option is to treat the parameters the same way one treats time: the first paper based on this idea was (Nouy, 2007). PGD, like any technique based on model reduction, can lead to the computation of numerous integrals, which can become very expensive. The Reference Point Technique (RPT), introduced in (Ladevèze, 1997) and currently undergoing further development, could lead to a drastic reduction in this cost (Ladevèze et al., 2009; Néron and Ladevèze, 2012).

Small-displacement (visco)plasticity problems. The book (Ladevèze, 1996, 1999) presents the state of the art prior to 1996. Much effort was devoted to modeling complex material behavior in a way which would be suitable for cyclic (visco)plasticity (Boisse et al., 1990; Cognard et al., 1999; Ladevèze et al., 1999). A two-time-scale approach for complex loading was introduced in (Ladevèze, 1991; Cognard and Ladevèze, 1993). Particular attention was paid to the direct calculation of the limit cycle (Ladevèze and

Rougée, 1985).

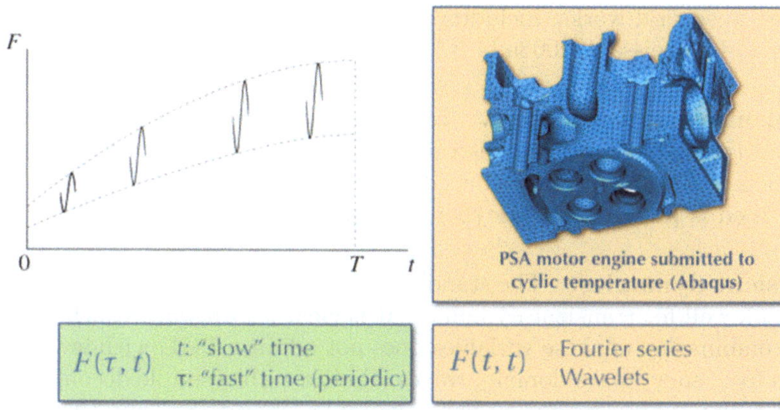

Figure 2. Two time-scale description

An application to a PSA combustion engine using an ABAQUS routine was presented in (Comte et al., 2006). A wavelet-based approach was introduced in (Néron, 2011). As mentioned previously, a rather robust PGD computational technique, the fruit of many years of research, is now available (Ladevèze et al., 2010) and will be described in detail in this paper.

Large-displacement and large-strain problems. A first approach using standard time-space variable separation was developed in (Bussy et al., 1990). This technique can be implemented easily, but is not very efficient in terms of the number of PGD modes necessary. The large-strain theory proposed in (Ladevèze, 1996, 1999) was applied to buckling problems, including post-buckling, in (Boucard et al., 1997) and to elastomer structures in (Aubard et al., 2002). These issues are being revisited today.

The multiscale PGD/LATIN method. This extension of the standard PGD method is derived from a mixed domain decomposition method which was introduced in (Ladevèze, 1985b; Ladevèze and Maurel, 1990). The present and final version (Ladevèze et al., 2002; Ladevèze and Nouy, 2003) is the end result of several works (Ladevèze and Lorong, 1992; Ladevèze and Dureisseix, 1999; Ladevèze et al., 2001, 2002). A detailed presentation of this final version is given here.

The multiphysics PGD/LATIN method. Here, the idea consists in introducing an abstract interface between physics which plays the same role as the usual interface material (Dureisseix et al., 2003b). This approach has led to several works, including particularly (Dureisseix et al., 2003b,a; Néron and Dureisseix, 2008a,b; Néron and Ladevèze, 2010).

Problems with parameters. Today, it is possible to use the PGD/LATIN method to solve engineering viscoplastic problems with a small number of parameters, and then build reduced models (Relun et al., 2013). The technique used is presented in (Heyberger et al., 2012, 2013).

Transient dynamics. The standard time/space variable separation does not work well for transient dynamics. It is clear that a wave which depends on a combination of these variables does not admit a PGD with few modes. In the frequency/space domain, waves admit a rather simple decomposition which is not a pure frequency/space variable separation.Using this decomposition, a PGD method has been introduced for the computation of large frequency band (Barbarulo et al., 2014).

Verification. Error indicators have been introduced in order to control the iterative process leading to a PGD approximation (Ladevze, 1996,99) A first tentative to derive an adaptive computational strategy has been developed in (Pelle and Ryckelynck, 2000). Recently, we have derived tools to get computable and accurate guaranteed error bounds for goal oriented quantities (Ladevèze and Chamoin, 2011, 2012) for linear elliptic and parabolic problems with parameters. They will be detailed in this paper. The extension to nonlinear problems is in progress.

Miscellaneous developments. The PGD/LATIN method has also been used for damage analysis in (Allix, 1992) and for inverse problems in (Allix and Vidal, 2002).

1.4 The simplest case: PGD of a given function

In this section, we follow (Ladevèze, 1996, 1999) and consider a time-space domain $I \times \Omega$ where $I = [0, T]$ and $\Omega \subset \mathbb{R}^d$. With f being a known scalar function defined over $I \times \Omega$, let us study the "best" p^{th}-order time-space approximation of function f related to a given norm:

$$\check{f}_p(t, \underline{M}) = \sum_{i=1}^{p} \lambda_i(t)\Lambda_i(\underline{M}) \tag{1}$$

One has:

$$\check{f}_p = \sum_{i=1}^{p} \lambda_i \Lambda_i = \arg \min_{\lambda_i, \Lambda_i} \| f - \sum_{i=1}^{p} \lambda_i \Lambda_i \|^2_{I \times \Omega} \tag{2}$$

λ_i and Λ_i constitute reduced bases of the spaces of the time and space functions respectively. If the norm chosen is that which is associated with the scalar product:

$$\langle f, g \rangle_{I \times \Omega} = \int_{I \times \Omega} fg \; d\Omega dt \tag{3}$$

the PGD of f is exactly the classical Singular Value Decomposition (Golub and Loan, 1996). One can prove that (2) is an eigenvalue problem whose eigenfunctions are the time functions λ_i. This problem can be rewritten as the stationarity of the Rayleigh quotient:

$$R(\lambda) = \frac{\| \langle f, \lambda \rangle_I \|^2_\Omega}{\| \lambda \|^2_I} \tag{4}$$

Under classical regularity assumptions on f, the eigenvalue problem has a countable sequence of eigensolutions $(\alpha_i^{-1}, \lambda_i)$, with the eigenvalues α_i^{-1} being positive and the eigenfunctions λ_i orthogonal.

Once the time functions λ_i have been determined, the corresponding space functions Λ_i are:

$$\Lambda_i = \frac{\langle f, \lambda_i \rangle_I}{\| \lambda_i \|^2_I} \tag{5}$$

The following convergence property is verified:

$$\| f - \check{f}_p \|_{I \times \Omega} \underset{p \to +\infty}{\longrightarrow} 0 \tag{6}$$

and a simple measure of the relative error is:

$$\eta_p = \frac{\| f - \check{f}_p \|_{I \times \Omega}}{\| \frac{1}{2}(f + \check{f}_p) \|_{I \times \Omega}} \tag{7}$$

These properties hold for other norms. To illustrate the relevance of separated time and space representations, let us consider the case of a randomly-obtained irregular function f defined over a time-space domain $I \times [0, L]$ in which space is one-dimensional. Figure 3 shows function f along with its first-, second- and third-order approximations. The relative error achieved with only 3 radial functions was less than 1%, which gives an idea of the remarkable accuracy achievable with the proposed time-space approximation.

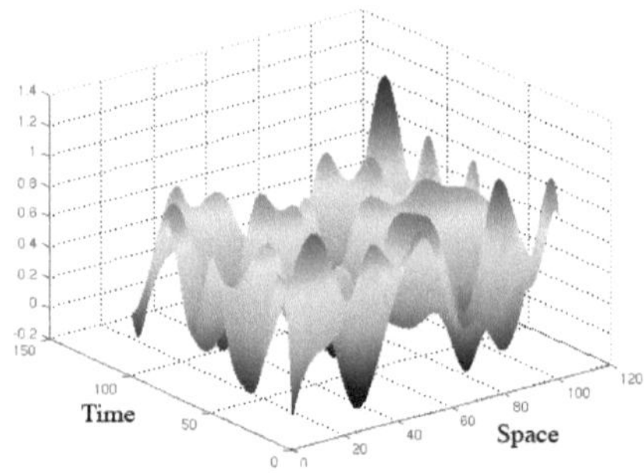

(a) Irregular time-space function f

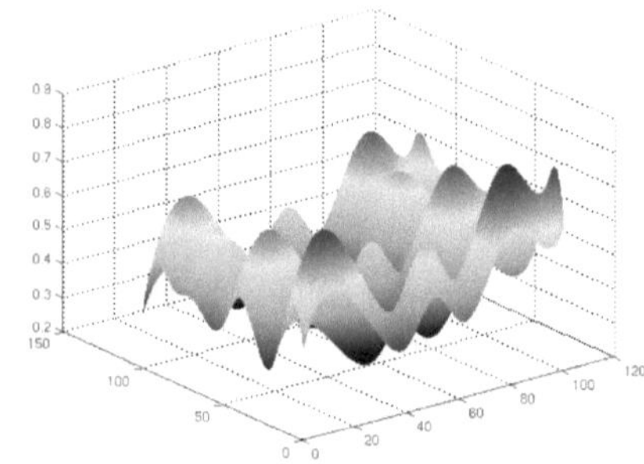

(b) First-order approximation $\check{f}_1$: error $\eta_1 = 3.9\%$

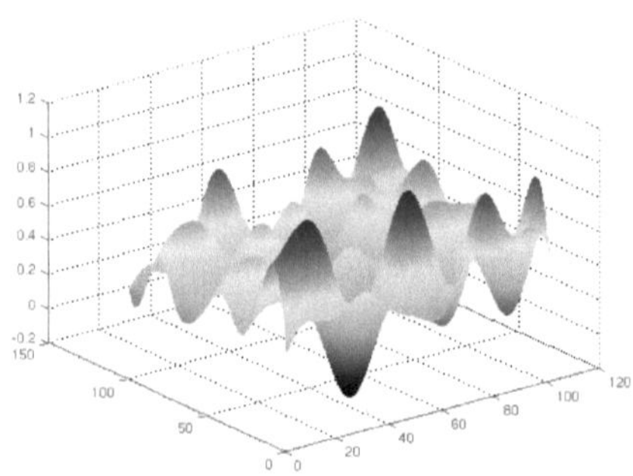

(c) Second-order approximation $\check{f}_2$: error $\eta_2 = 1.5\%$

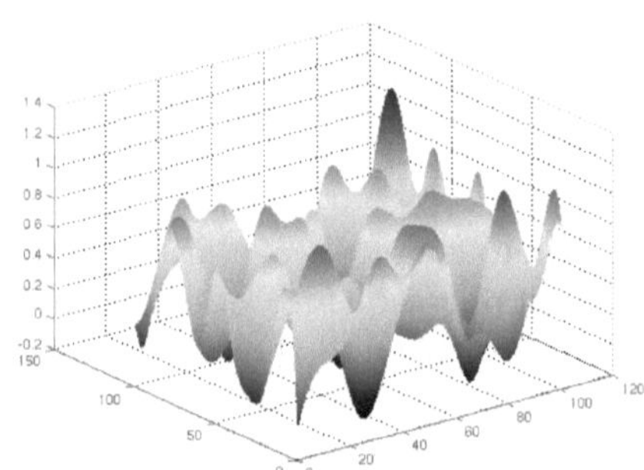

(d) Third-order approximation $\check{f}_3$: error $\eta_3 = 0.6\%$

Figure 3. Time-space approximations of an irregular function f

2 PGD-verification for linear elliptic and parabolic problems

2.1 Introduction

Performances of PGD-reduced models have been shown qualitatively in many applications exhibiting variations in loading, boundary or initial conditions, material characteristics, geometry, ..., even for nonlinear problems. That is not enough: a major constraint for their transfer and intensive use in industry is to be sure of their accuracy. Indeed, certifying the accuracy of the PGD solution is a fundamental issue in order to perform robust and reliable design. This control requires mastering the number of PGD modes which are computed, but also the numerical methods which are performed

in these computations.
There are actually very few works which have addressed the control of PGD-based approximations until now. Basic results on a priori error estimations for representations with separation of variables are given in (Ladevèze, 1996, 1999), whereas error indicators mostly devoted to adaptivity can be found in (Pelle and Ryckelynck, 2000; Ammar et al., 2010). A first robust approach for PGD verification, using the concept of Constitutive Relation Error (CRE) (Ladevèze and Pelle, 2004; Ladevèze, 2008), was proposed recently (Ladevèze and Chamoin, 2011; Chamoin and Ladevèze, 2012; Ladevèze and Chamoin, 2012). It applies to linear elliptic or parabolic problems depending on parameters, and provides for guaranteed PGD-reduced models, both on global error and error on specific outputs on interest. Furthermore, the approach enables to assess contributions of various error sources (space and time discretizations, truncation of the PGD decomposition, etc.), which can help driving adaptive strategies. These works are reported here. Let us also quote works concerning other reduced model methods (Grepl and Patera, 2005; Rovas et al., 2006; Boyaval et al., 2009).

In order to get guaranteed error bounds there is only one way which can be described with different words. Here, we use the Constitutive Relation Error method developed during years in particular at LMT-Cachan (see (Ladevèze and Pelle, 2004)). An extension to nonlinear problems such as (visco)-plasticity ones, even in dynamics, was recently proposed under the small displacement assumption (Ladevèze, 2008). The first application to plasticity in quasi-static was done in (Ladevèze et al., 2012). In dynamic for visco-elastic materials, we have given the first guaranteed error bounds in (Waeyetens et al., 2012).
To enter in the CRE-machinery, one needs an approximation which satisfies the kinematic constraints and the equilibrium equations in the finite element sense. Then, using an equilibration technique, one builds a new approximation which satisfies all the equations except the constitutive relations. The approximation quality is thus defined by Constitutive Relation Errors (Ladevèze and Maunder, 1996; Ladevèze and Pelle, 2004; Ladevèze et al., 2010; Pled et al., 2011). The main and only difficulty with the PGD-approximations is that they do not satisfy the FE equilibrium. Then, to overcome this difficulty, we have proposed a rather simple technique which associates to the PGD-approximation and the data a new approximation able to enter in the CRE-machinery. This important point is detailed here.

2.2 The reference problem

We consider a transient diffusion problem defined on an open bounded domain $\Omega \subset \mathbb{R}^d$ ($d = 1, 2, 3$), with boundary $\partial\Omega$, over a time interval $\mathcal{I} = [0, T]$. We assume that a prescribed zero temperature is applied on part $\partial_u\Omega \neq \emptyset$ of $\partial\Omega$ and that the domain is subjected to a time-dependent thermal loading that consists of: (i) a given thermal flux $r_d(\mathbf{x}, t)$ on $\partial_q\Omega \subset \partial\Omega$, with $\partial_u\Omega \cap \partial_q\Omega = \emptyset$ and $\overline{\partial_u\Omega \cup \partial_q\Omega} = \partial\Omega$; (ii) a source term $f_d(\mathbf{x}, t)$ in Ω.

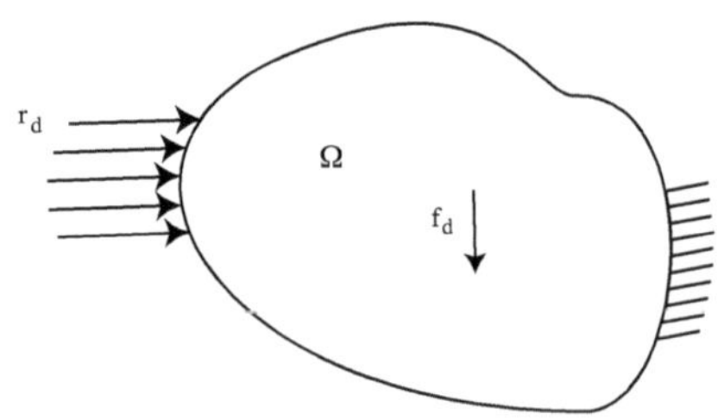

Figure 4. Representation of the reference problem.

For the sake of simplicity, we consider that initial conditions are set to zero. The material that composes Ω is assumed to be isotropic but heterogeneous and partially unknown. Therefore, diffusion coefficient μ and thermal capacity c depend on space variable $\mathbf{x}$ but also on a set of N parameters $\mathbf{p} = [p_1, p_2, \dots, p_N]$ belonging to a given bounded domain $\Theta = \Theta_1 \times \Theta_2 \times \dots \times \Theta_N$.

The associated mathematical problem consists of finding the temperature-flux pair $(u(\mathbf{x}, t, \mathbf{p}), \varphi(\mathbf{x}, t, \mathbf{p}))$, with $(\mathbf{x}, t, \mathbf{p}) \in \Omega \times \mathcal{I} \times \Theta$, that verifies:

- the thermal constraints:

$$u = 0 \quad \text{in } \partial_u\Omega \times \mathcal{I} \times \Theta \tag{8}$$

- the equilibrium equations:

$$c\frac{\partial u}{\partial t} = -\boldsymbol{\nabla}\cdot\varphi + f_d \quad \text{in } \Omega \times \mathcal{I} \times \Theta \quad ; \quad \varphi\cdot\mathbf{n} = r_d \quad \text{in } \partial_q\Omega \times \mathcal{I} \times \Theta \tag{9}$$

- the constitutive relation:

$$\varphi = -\mu\boldsymbol{\nabla}u \quad \forall(\mathbf{x}, t, \mathbf{p}) \in \Omega \times \mathcal{I} \times \Theta \tag{10}$$

- the initial conditions:

$$u_{|t=0^+} = 0 \quad \forall(\mathbf{x}, \mathbf{p}) \in \Omega \times \Theta \tag{11}$$

$\mathbf{n}$ denotes the outgoing normal to Ω. In the following, in order to be consistent with other linear problems encountered in Mechanics (linear elasticity for instance), we carry out the change of variable $\boldsymbol{\varphi} \to -\boldsymbol{\varphi}$ which leads, in particular, to the new constitutive relation $\boldsymbol{\varphi} = \mu \boldsymbol{\nabla} u$.
Defining $\mathcal{V} = H_0^1(\Omega) = \{v \in H^1(\Omega), v_{|\partial_u\Omega} = 0\}$, the weak formulation in space of the previous problem reads for all $(t, \mathbf{p}) \in \mathcal{I} \times \Theta$:

$$\text{Find } u(\mathbf{x}, t, \mathbf{p}) \in \mathcal{V} \text{ such that } b(u, v) = l(v) \quad \forall v \in \mathcal{V} \tag{12}$$

with $u_{|t=0^+} = 0$. Bilinear form $b(\bullet, \bullet)$ and linear form $l(\bullet)$ are defined as:

$$b(u,v) = \int_\Omega \left\{ c\frac{\partial u}{\partial t} v + \mu \boldsymbol{\nabla} u \cdot \boldsymbol{\nabla} v \right\} \mathrm{d}\Omega \quad ; \quad l(v) = \int_\Omega f_d v \mathrm{d}\Omega - \int_{\partial_q\Omega} r_d v \mathrm{d}S \tag{13}$$

As regards the full weak formulation, we introduce the functional spaces $\mathcal{T} = L^2(\mathcal{I})$, $\mathcal{P}_i = L^2(\Theta_i)$, and $L^2(\mathcal{I}, \Theta; \mathcal{V}) = \mathcal{V} \otimes \mathcal{T} \otimes_{n=1}^N \mathcal{P}_n$. We therefore search solution $u \in L^2(\mathcal{I}, \Theta; \mathcal{V})$, with $\dfrac{\partial u}{\partial t} \in L^2(\mathcal{I}, \Theta; L^2(\Omega))$, such that:

$$B(u, v) = L(v) \quad \forall v \in L^2(\mathcal{I}, \Theta; \mathcal{V}) \tag{14}$$

with

$$\begin{aligned} B(u,v) &= \int_\Theta \left[\int_\mathcal{I} b(u,v)\mathrm{d}t + \int_\Omega cu(\mathbf{x},0^+)v(\mathbf{x},0^+)\mathrm{d}\Omega \right] \mathrm{d}\mathbf{p} \\ L(v) &= \int_\Theta \int_\mathcal{I} l(v)\mathrm{d}t\mathrm{d}\mathbf{p} \end{aligned} \tag{15}$$

The exact solution of (14), which is usually out of reach, is denoted u_{ex} (and $\boldsymbol{\varphi}_{ex} = \mu \boldsymbol{\nabla} u_{ex}$). It is classically approximated using the FEM in space associated with a time integration scheme and a given grid in the parameter space Θ.

<u>Remark 1</u>: in the steady-state case, we merely consider:

$$b(u,v) = \int_\Omega \mu \boldsymbol{\nabla} u \cdot \boldsymbol{\nabla} v \, d\Omega \quad ; \quad B(u,v) = \int_\Theta b(u,v)\, d\mathbf{p} \quad ; \quad L(v) = \int_\Theta l(v)\, d\mathbf{p} \tag{16}$$

2.3 Computation of the PGD approximation: principle

For the sake of simplicity, let us consider the case where parameters have given values. In the following, we present a classical version of the PGD technique, called *progressive Galerkin computational technique* and inspired

from fixed-point algorithms used to solve eigenvalue problems.
We assume that a PGD approximation of order $m-1$ has been computed. The PGD approximation of order m is then defined as:

$$u(\mathbf{x},t) \approx u_m(\mathbf{x},t) = \sum_i^m \psi_i(\mathbf{x})\lambda_i(t) = \mathbf{\Psi}_m(\mathbf{x}) \cdot \mathbf{\Lambda}_m(t) \tag{17}$$

where $\mathbf{\Psi}_m(\mathbf{x}) = [\psi_1(\mathbf{x}), \psi_2(\mathbf{x}), \ldots, \psi_m(\mathbf{x})]^T$ (resp. $\mathbf{\Lambda}_m(t) = [\lambda_1(t), \lambda_2(t), ..., \lambda_m(t)]^T$) is a reduced basis of space (resp. time) functions, and $m \in \mathbb{N}$ is the representation order. In the PGD approach, neither functions $\lambda_i(t)$ nor functions $\psi_i(\mathbf{x})$ are given initially, and both families are computed on the fly.

A new couple $(\psi, \lambda) \in \mathcal{V} \times \mathcal{T}$ is defined for the order m representation as the one that verifies the following double Galerkin orthogonality criterion:

$$B(u_{m-1} + \psi\lambda, \psi\lambda^* + \psi^*\lambda) = L(\psi\lambda^* + \psi^*\lambda) \quad \forall \lambda^* \in \mathcal{T}, \forall \psi^* \in \mathcal{V} \tag{18}$$

We can thus define the two following applications:

- $S_m : \mathcal{T} \rightarrow \mathcal{V}$ is the application that maps a time function λ into a space function $\psi = S_m(\lambda)$ defined as:

$$B(u_{m-1} + \psi\lambda, \psi^*\lambda) = L(\psi^*\lambda) \quad \forall \psi^* \in \mathcal{V} \tag{19}$$

  It is associated to a space problem;
- $T_m : \mathcal{V} \rightarrow \mathcal{T}$ is the application that maps a space function ψ into a time function $\lambda = T_m(\psi)$ defined as:

$$B(u_{m-1} + \psi\lambda, \psi\lambda^*) = L(\psi\lambda^*) \quad \forall \lambda^* \in \mathcal{T} \tag{20}$$

  It is associated to a time problem (scalar ODE).

A couple (ψ, λ) verifies (18) if and only if $\psi = S_m(\lambda)$ and $\lambda = T_m(\psi)$, which is a nonlinear problem. This version of PGD can then be interpreted as a pseudo eigenvalue problem, functions ψ and λ respectively being dominating eigenfunctions of applications $G_m := S_m \circ T_m$ and $\widetilde{G}_m := T_m \circ S_m$ (Nouy, 2010). This interpretation is useful as it enables to propose dedicated algorithms, inspired from those used in eigenvalue problems, to build the PGD representation. One of these algorithms, denoted *power iterations algorithm*, is described below:

→ **for** $m = 1$ to m_{max}
→ define $\lambda^{(0)}$ (initialization)
→ **for** $k = 1$ to k_{max}
→ compute $\psi^{(k)} = S_m(\lambda^{(k-1)})$
→ normalize $\psi^{(k)}$ ($||\psi^{(k)}||_\Omega = 1$)

→ compute $\lambda^{(k)} = T_m(\psi^{(k)})$
→ check convergence of $(\psi^{(k)}\lambda^{(k)})$
→ **end for**
→ define $\psi_m = \psi^{(k)}$ and $\lambda_m = \lambda^{(k)}$
→ define $u_m = u_{m-1} + \psi_m \lambda_m$ and check convergence
→ **end for**

We therefore remark that u_m is obtained from the solution of a few space or time problems, without any knowledge on u; the first step $\psi^{(k)} = S_m(\lambda^{(k-1)})$ of the iterative strategy, which is the most costly, consists in solving a space problem with fixed time function $\lambda^{(k-1)}$; the second step $\lambda^{(k)} = T_m(\psi^{(k)})$ consists in solving a time problem over the whole domain $\mathcal{I}$, with fixed space function $\psi^{(k)}$. In practice, both problems are solved with classical numerical methods, i.e. the Finite Element Method (FEM) in space and time. The time problem could also be solved through a first order differential equation.

2.4 The proposed PGD-computation technique

The main lines of this technique were defined around 1990. The idea is to keep, at iteration m, the $(m-1)$ shape functions depending on the space variable because they are the most costly to compute. So, at iteration m, there is a first stage named very often "preliminary stage" which is in fact a POD step. From available space functions $(\psi_1,\psi_2,\ldots,\psi_{m-1})$, iteration m then consists of two steps:

1. POD step.
   - Find $\lambda_i^m(t)$ $(1 \leq i \leq m-1)$ such that

$$B\Big(\sum_{i=1}^{m-1} \lambda_i^m(t)\psi_i, \psi_j\lambda^*(t)\Big) = L(\psi_j\lambda^*(t)) \quad \forall\lambda^* \in \mathcal{T}, 1 \leq j \leq m-1 \tag{21}$$

   One gets a time problem which can be solved as a F.E. one. Another way is to discretize the differential equations obtained from the continuum mechanics framework.
   - Stop if the error $\leq \varepsilon_0$
2. PGD step

   One adds λ_m^m and ψ_m using the progressive Galerkin technique for which one stops after few iterations, practically between 2 and 4. For the following illustrations, we stopped after 4 iterations. The approximation is then $\sum_{i=1}^m \lambda_i^m \psi_i$. In fact, the used approximation is the one we get when the error is less than ε_0.

Remark 2: It could be interesting to orthonormalize functions ψ_i.

Illustration

In the following numerical experiment, the power iterations algorithm is initialized with $\lambda^{(0)}(t) = t$. We use the FEM with linear elements to solve space problems, and a forward Euler scheme to solve time problems. The number of iterations performed in the power iterations algorithm is fixed to $k_{max} = 4$.

We consider the 2D structure of Fig. 5, which contains two rectangular holes in which a fluid circulates; using symmetries, we study one quarter (denoted Ω) of the full 2D domain. It is subjected over $[0, T]$ to a given flux $q_d = -1$ applied on the hole boundary; a source term of the form $f_d(x, y) = 200xy$ is also applied in Ω. For the numerical study, we choose $T = 10$, $\mu = 1$, and $c = 1$. The space mesh used to discretize Ω consists of $N_e = 50$ quadrangular elements, and time domain $[0, T]$ is divided into $N_p = 1000$ time steps.

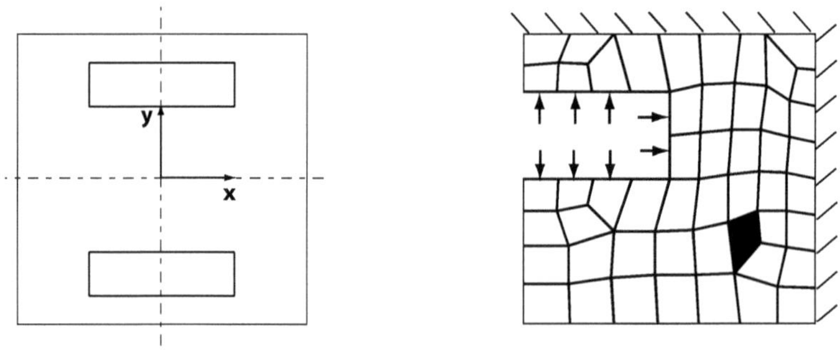

Figure 5. 2D domain considered (left), and associated FE mesh (right).

We represent in Fig. 6 first modes (ψ_m, λ_m) obtained for $m_{max} = 5$.

2.5 A main technical point: construction of a FE-equilibrated PGD approximation

To enter in the CRE-machinery, the considered approximation should satisfy the FE equilibrium; that is mandatory. Then there is a main difficulty with PGD approximations: the returned approximation $u_m = \sum_{i=1}^{m} \lambda_i^m \psi_i$ (or $v_m = \sum_{i=1}^{m} \lambda_i^{m+1} \psi_i$) does not satisfy the FE equilibrium. Here, a new PGD approximation satisfying the FE equilibrium is derived from u_m and the data. The idea is to use the following properties of u_i for

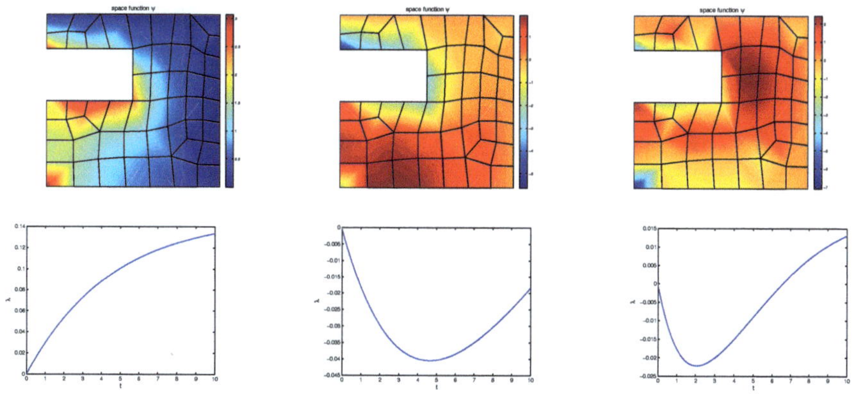

Figure 6. Space functions ψ_m (top) and time functions λ_m (bottom) for $m = 1, 2, 3$ (from left to right).

$1 \leq i \leq m$ when the power iterations algorithm is stopped after a space problem:

$$B(u_i, \lambda_i^i \psi^*) = L(\lambda_i^i \psi^*) \quad \forall \psi^* \in \mathcal{V} \tag{22}$$

Let us first consider the second member of (22) which is supposed to be equal to:

$$L = \sum_{j=1}^{n} \gamma_j(t) \bar{L}_j \tag{23}$$

After solving several steady-state problems, one gets $\mathbf{r}_d = \sum_{j=1}^{n} \gamma_j \mathbf{r}_{j,d}$ such that:

$$\int_0^T \int_\Omega \mathbf{r}_d \cdot \boldsymbol{\nabla} \psi^* \mathrm{d}\Omega \mathrm{d}t = L(\lambda_i^i \psi^*) \quad \forall \psi^* \in \mathcal{V} \tag{24}$$

If $L(a^*) = 0$ for a^* constant, the procedure should be slightly modified.

The fundamental relation (22) can be rewritten as:

$$\int_\Omega (\sum_{j=1}^{i} A_j^i \psi_j) \psi^* \mathrm{d}\Omega + \int_\Omega \mathbf{Q}_i \cdot \boldsymbol{\nabla} \psi^* \mathrm{d}\Omega = 0 \quad \forall \psi^* \in \mathcal{V} \tag{25}$$

with $A_j^i = \int_0^T c \dot{\lambda}_j^i \lambda_i^i \mathrm{d}t$ and $\mathbf{Q}_i = \int_0^T (k \boldsymbol{\nabla} u_i \lambda_i^i - \mathbf{r}_d) \mathrm{d}t$, for $1 \leq i \leq m$. Finally:

$$\int_\Omega \mathbb{A} \boldsymbol{\Psi}_m \psi^* \mathrm{d}\Omega + \int_\Omega \mathbf{Q} \cdot \boldsymbol{\nabla} \psi^* \mathrm{d}\Omega \quad \forall \psi^* \in \mathcal{V} \tag{26}$$

with $\mathbf{Q} = [\mathbf{Q}_1, \mathbf{Q}_2, \ldots, \mathbf{Q}_m]^T$. Supposing the material constant c is uniform over the domain Ω, one gets a constant matrix $\mathbb{A}$. It follows for any t belonging to $[0, T]$:

$$\int_\Omega c\dot{\mathbf{\Lambda}}_m^m \cdot \mathbf{\Psi}_m \psi^* \mathrm{d}\Omega + \int_\Omega c(\dot{\mathbf{\Lambda}}_m^m \cdot \mathbb{A}^{-1}\mathbf{Q}) \cdot \mathbf{\nabla}\psi^* \mathrm{d}\Omega = 0 \quad \forall \psi^* \in \mathcal{V} \tag{27}$$

and consequently $(u_m, c\dot{\mathbf{\Lambda}}_m^m \cdot \mathbb{A}^{-1}\mathbf{Q} + \mathbf{r}_d)$ satisfies the FE equilibrium at any time belonging to $[0, T]$. One notes $\hat{u} = u_m$ and $\bar{\bar{\varphi}} = c\dot{\mathbf{\Lambda}}_m^m \cdot \mathbb{A}^{-1}\mathbf{Q} + \mathbf{r}_d$.

Remark 3: To satisfy only the discretized problem, a better PGD approximation could be built when a time FE approach is used. Let $N_i(t)$, $1 \leq i \leq k$, be the time functions related to the discretization over $[0, T]$ and $\mathbf{N}^T = [N_1, \ldots, N_k]$. A PGD approximation which satisfies the discretized equilibrium equation and the kinematic constraints is $\hat{u}$ and $\bar{\varphi} = \mathbf{N}^T (\int_0^T [\mathbf{N}\mathbf{N}^T])^{-1}\mathbf{R}$ where $\mathbf{R}^T = [\mathbf{R}_1 \ldots \mathbf{R}_k]$ and $\mathbf{R}_i = \int_0^T \bar{\bar{\varphi}} N_i \, dt$.

Remark 4: The techniques to get $\bar{\varphi}$ and $\bar{\bar{\varphi}}$ can be extended to the case where the constant $c(t, \mathbf{x})$ can be written as $c = \sum_{i=1}^{\ell} \alpha_i(t)\beta_i(\mathbf{x})$.

Remark 5: If one uses a discretized differential equation for solving the time problem, it is possible to get an equivalent variational formulation introducing weight functions (Zienkiewicz and Taylor, 2000). Then one can use the principle of the technique developed in Remark 3 to get a good approximation satisfying the discretized differential equation.

2.6 The CRE method - global errors

The Constitutive Relation Error method: principle. The CRE machinery starts with a solution which is FE admissible. Using an equilibration technique (see (Ladevèze and Leguillon, 1983; Ladevèze and Maunder, 1996; Ladevèze and Pelle, 2004; Ladevèze et al., 2010; Pled et al., 2011, 2012)) one builds an admissible solution $(\hat{u}, \hat{\varphi})$ of the problem for nearly incompressible materials, one has (Ladevèze et al., 1992). The CRE measure, that depends on $\mathbf{p}$, then reads:

$$E_{CRE}^2(\mathbf{p}) = \frac{1}{2} \int_\mathcal{I} \int_\Omega \frac{1}{k} [\hat{\varphi} - \mu \mathbf{\nabla} \hat{u}] \cdot [\hat{\varphi} - \mu \mathbf{\nabla} \hat{u}] \mathrm{d}\Omega \mathrm{d}t \equiv \frac{1}{2} |||\hat{\varphi} - \mu \mathbf{\nabla} \hat{u}|||^2 \tag{28}$$

where $||| \bullet |||$ is an energy norm in the space-time domain, and one has the extension of the Prager-Synge theorem:

$$|||\varphi_{ex} - \hat{\varphi}^*|||^2 + \frac{1}{2} \int_\Omega c(u_{ex} - \hat{u})^2_{|T} \mathrm{d}\Omega = \frac{1}{2} E_{CRE}^2 \tag{29}$$

with $\hat{\varphi}^* = \frac{1}{2}[\hat{\varphi} + \mu \nabla \hat{u}]$. The value of E_{CRE} can be used as an estimate of the global error between u_{ex} and u_m.

Remark 6: *Again, in the steady-state case, we would consider:*

$$E^2_{CRE}(\mathbf{p}) = \frac{1}{2} \int_\Omega \frac{1}{\mu}[\hat{\varphi} - \mu \nabla \hat{u}] \cdot [\hat{\varphi} - \mu \nabla \hat{u}] d\Omega \equiv \frac{1}{2} ||\hat{\varphi} - \mu \nabla \hat{u}||^2 \tag{30}$$

$$||\varphi_{ex} - \hat{\varphi}^*||^2 = \frac{1}{2} E^2_{CRE} \tag{31}$$

Specific error indicators. Specific error indicators are needed to control the computation process. The accuracy of the PGD approximation depends both on the discretization and on the number of computed PGD modes. The idea to define such indicators, which is not restricted to the PGD framework, has been applied in other works and is quite simple.

For the PGD error indicator, one considers as the reference the discretized problem. Using still the CRE-method, one gets:

$$E^2_{PGD} = \frac{1}{2} \int_0^T \mathrm{d}t \int_\Omega \frac{1}{\mu} (\bar{\varphi} - \mu \nabla \hat{u}) \cdot (\bar{\varphi} - \mu \nabla \hat{u}) \tag{32}$$

The error indicator which is specific to the discretization is defined by:

$$E^2_{dis} = E^2_{CRE} - E^2_{PGD} \tag{33}$$

and one has specific error indicators related to the mesh and time discretization:

$$E^2_{dis} = E^2_{mesh} + E^2_{dis,t} \tag{34}$$

with $E^2_{mesh} = \frac{1}{2} \int_0^T \mathrm{d}t \int_\Omega \frac{1}{\mu} (\hat{\varphi} - \bar{\bar{\varphi}}) \cdot (\hat{\varphi} - \bar{\bar{\varphi}})$.

Illustration

In Fig. 9, the value of the relative error estimate $\bar{E}_{CRE} = E_{CRE}/|||\hat{\varphi}|||_{\mu^{-1}}$ is given with respect to the PGD order m which is used ($m \in [1, 10]$). We observe that the error estimate reaches convergence for $m = 4$. Let us note that for this value of m, the CPU time required to obtain the error estimate (normalized by the time necessary to solve the reference problem using PGD) is 2.8.

The values of three estimates E_{CRE}, E_{PGD}, and E_{dis} are plotted in Fig. 9 with respect to order m. We observe that indicator E^2_{dis} provides for

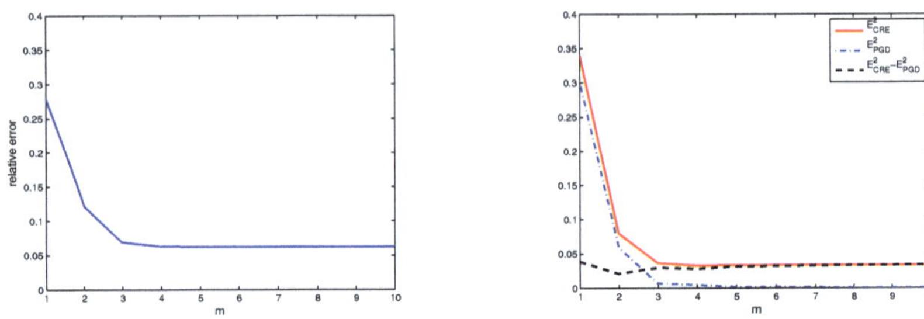

Figure 7. Evolution of estimates E^2_{CRE}, E^2_{PGD}, and E^2_{dis} with respect to PGD order m.

a relevant assessment of the discretization error, even for small values of m. Furthermore, we can see that indicator E_{dis} becomes larger than E_{PGD} for $m \geq 3$; this could be used in an optimized adaptive process in which the number of computed modes would be first increased, up to $m = 3$, before refining time and space discretizations in order to improve the quality of the PGD solution.

2.7 The CRE method - goal-oriented error estimation

Guaranteed bounding. Let I be an output of interest defined by the extraction pair $(\boldsymbol{\varphi}_\Sigma, \mathbf{f}_\Sigma)$:

$$I(\mathbf{p}) = \int_{\mathcal{I}} \int_{\Omega} \{\boldsymbol{\varphi}_\Sigma \cdot \boldsymbol{\nabla} u + \mathbf{f}_\Sigma \cdot u\} \, \mathrm{d}\Omega \mathrm{d}t \tag{35}$$

$\boldsymbol{\varphi}_\Sigma(\mathbf{x}, t)$ and $\mathbf{f}_\Sigma(\mathbf{x}, t)$ can possibly be Dirac distributions. We therefore introduce the associated adjoint problem, and compute an approximate (resp. admissible) PGD solution $(\tilde{u}_{\tilde{m}}, \tilde{\boldsymbol{\varphi}}_{\tilde{m}})$ (resp. $(\hat{\tilde{u}}_{\tilde{m}}, \hat{\tilde{\boldsymbol{\varphi}}}_{\tilde{m}})$) for this problem. In practice, the PGD solution of the adjoint problem is performed using an order $\tilde{m}$ possibly different from m, as well as introducing local enrichment functions in the vicinity of the space-time region where I is defined (Ladevèze et al., 2010).

The fundamental result leading to guaranteed and accurate error upper bounds is given in (Ladevèze, 2006) with developments for linear elliptic and parabolic problems (see (Chamoin and Ladevèze, 2007; Ladevèze, 2008; Chamoin and Ladevèze, 2008; Ladevèze and Chamoin, 2010; Ladevèze et al., 2013)):

$$|I_{ex} - I_m - I_{corr}| \leq E_{CRE} \tilde{E}_{CRE} \tag{36}$$

where $I_{ex}(\mathbf{p})$ (resp. $I_m(\mathbf{p})$) is the exact (resp. approximated by PGD) value of the output of interest, $I_{corr}(\mathbf{p})$ is a correction term computed from approximate solutions of both reference and adjoint problems, and $E_{CRE}(\mathbf{p})$ (resp. $\tilde{E}_{CRE}(\mathbf{p})$) is the constitutive relation error of the reference (resp. adjoint) problem. Let us note that this bounding result does not use any Galerkin orthogonality property, but only properties of admissible solutions.
Consequently, strict bounds on the local error $I_{ex} - I_m$ (or directly on I_{ex}) can be obtained for any value $\mathbf{p}$ of material parameters. The second member of the fundamental result (36) is obtained thanks to the Cauchy Schwartz inequality. A better upper bound is given in (Ladevèze et al., 2013).

As a simple example, we consider the 2D structure of Figure 8 which is a section presenting two rectangular holes in which a fluid circulates. It is subjected to a given source term $f_d(x,y) = 200xy$, a given flux $r_d(t) = -1$ on the holes boundaries, and a zero temperature is imposed on other boundaries. Using symmetries, we only keep the upper right quarter of the structure that we denote Ω. We consider that the diffusion coefficient k is fluctu-

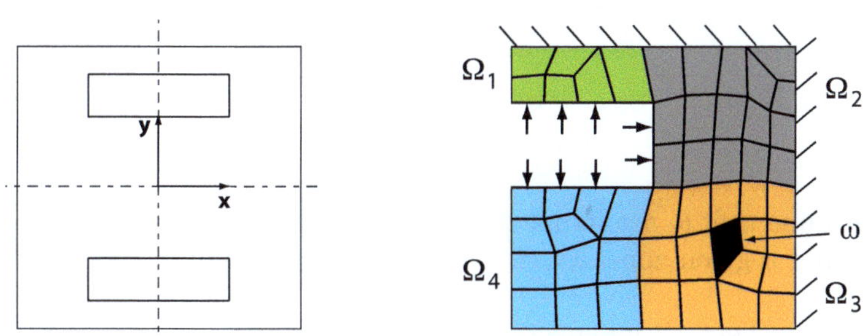

Figure 8. The 2D reference problem.

ating but remains piecewise homogeneous, i.e. it is homogeneous in each of the four non-overlapping subdomains Ω_i $(i = 1,2,3,4)$ defined in Figure 8 and such that $\overline{\Omega_1 \cup \Omega_2 \cup \Omega_3 \cup \Omega_4} = \Omega$. Furthermore, the thermal capacity c is supposed homogeneous in the whole domain Ω. In the following, the two material coefficients are defined as:

$$\mu(\mathbf{x},\theta_i) = 1 + \sum_{i=1}^{4} g_i I_{\Omega_i}(\mathbf{x})\theta_i \qquad c(\mathbf{x},\theta_5) = 1 + 0.2\,\theta_5 \tag{37}$$

with $[g_1, g_2, g_3, g_4] = [0.1, 0.1, 0.2, 0.05]$, $\theta_i \in [-2,2]$, and $I_{\Omega_i}(\mathbf{x})$ denoting the indicatrix function of subdomain Ω_i. The resulting solution $u(\mathbf{x}, t, \theta_1, \theta_2, \theta_3, \theta_4, \theta_5)$ is searched using the PGD technique, with four-nodes quadrangular elements in space and a forward Euler time scheme.

Figure 9 gives the evolution of the constitutive relation error (seen as a global error estimate) with respect to the number m of PGD modes taken in the approximation; this estimate is computed for $\theta_i = 0$ $(i = 1, \ldots, 5)$. We observe that after 5 modes, the error reaches an asymptotic value that corresponds to the discretization error.

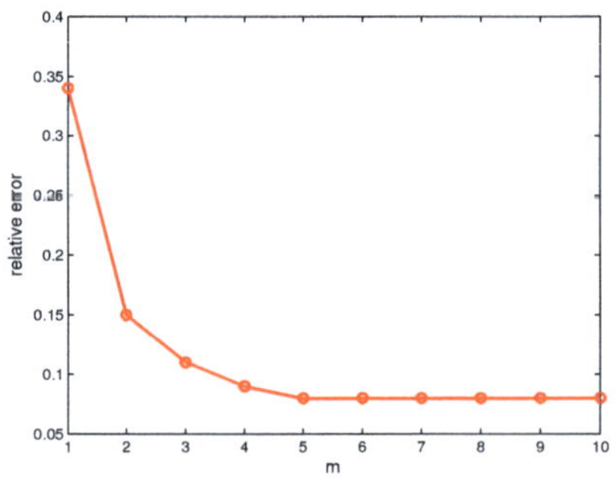

Figure 9. Relative error estimate with respect to the number m of PGD modes.

Assuming that θ_i are (truncated) centered reduced normal variables, and considering a given zone $\omega \subset \Omega$ (see Figure 8), we study two quantities of interest:

- the mathematical expectation (in the probabilistic sense) of the mean value of u inside ω at final time T:

$$I_1 = \mathrm{E}\left[\frac{1}{|\omega|}\int_\omega u_{|T}\mathrm{d}\Omega\right] \tag{38}$$

- the maximal value of the mean value of u inside ω at final time T:

$$I_2 = \sup_{\theta_i}\frac{1}{|\omega|}\int_\omega u_{|T}\mathrm{d}\Omega \tag{39}$$

We choose an order $m = 3$ for the approximate PGD solution of the reference problem. For both I_1 and I_2, obtained normalized upper bounds on $I_{ex} - I_m - I_{corr}$ as well as specific error indicators are given in Figure 10 with respect to the number $\tilde{m}$ of computed PGD modes for the adjoint solution.

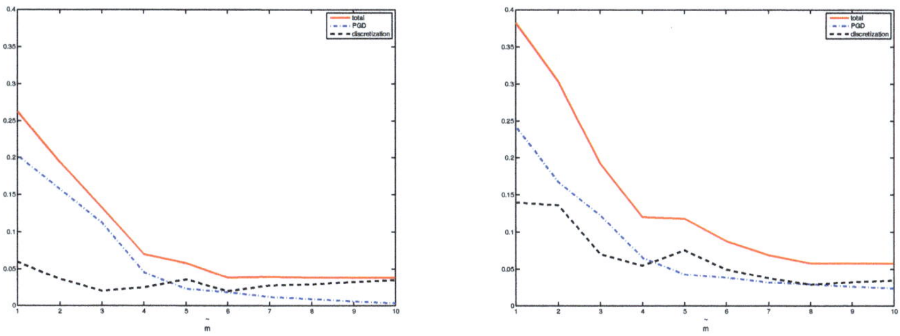

Figure 10. Normalized upper error bound and error indicators with respect to the number $\tilde{m}$ of PGD modes used for the adjoint solution: I_1 (left), I_2 (right).

We observe that the correcting term I_{corr} enables to assess I_{ex} very effectively. The remaining asymptotic error could be decreased by improving discretizations used to compute PGD modes.

2.8 Toward guaranteed virtual charts

An adaptive computational strategy could be easily derived with the error indicators related to the adjoint problem (Figure 11). The idea is to refine the discretization (mesh and time) when the convergence slope of the error (with respect to the number of modes) becomes relatively small. In (Ladevèze and Pelle, 2004), discretization tools are given to define the optimal discretization. One uses error indicators E_{mesh} and $E_{dis,t}$ and their local contributions. The asymptotic convergence rate is also taken into account.

2.9 Prospects

PGD-reduced models are a promising tool for solving complex engineering problems: few PGD modes are enough to get an accurate approximation certainly here for academic example. However, a central and a main question is to guarantee their accuracy. The verification method described here is an answer to this challenge for linear elliptic and parabolic problems (see (Chamoin et al., 2012) for engineering applications). It can be applied in the case of numerous parameters (such as stochastic problems).
The scientific challenge today is to derive PGD verification tools for time-

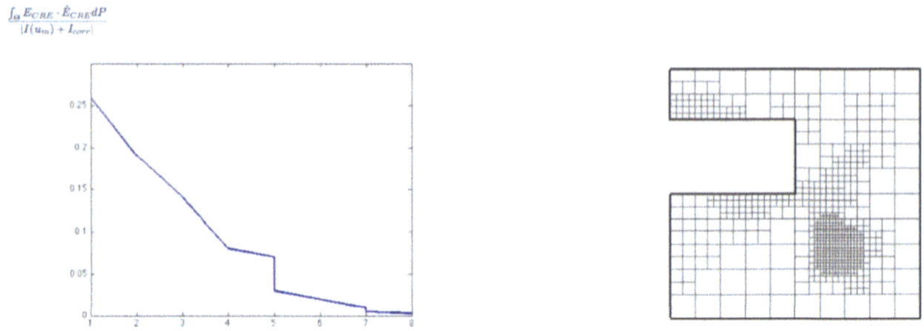

Figure 11. Convergence of the error on I with respect to the number of modes (left) and obtained refined mesh in space (right) using the adaptive strategy.

dependent nonlinear problems. The Constitutive Relation Error method seems to be a promising way with its verification tools gathered in book (Ladevèze and Pelle, 2004). However guaranteed error bounds are not available in general.

3 PGD for time dependent nonlinear problems

3.1 Introduction: the PGD/LATIN method in brief

The LATIN method addresses the key issue (or, at least, one of the key issues) of the computational strategy itself which is used to solve the large-scale, nonlinear and time-dependent problems. First, looking at classical techniques, one can make a common-sense remark: classical computational strategies appear to work for everything; therefore, like all things universal, they cannot achieve high performance for one domain in particular. There is no miracle: high performance requires custom-made techniques. Therefore, the LATIN method is designed as a mechanics-based computational strategy whose aim is to achieve the best possible performance level in mechanics. Consequently, this alternative computational approach is rooted in some remarkable properties which are verified by most of the models encountered in structural mechanics. This iterative strategy differs from classical incremental or step-by-step techniques in that, at each iteration, it produces an approximation of the complete structural response over the whole loading history being considered. In other words, the name LATIN (LArge Time Increment) was not chosen very well because the method is essentially nonincremental. What is important is the fact that the LATIN method is global over the time-space domain. The LATIN method was first introduced in (Ladevèze, 1985a) and in (Ladevèze, 1989) for material models with internal variables. A review of the state-of-the-art and more recent extensions can be found in the book (Ladevèze, 1996, 1999).

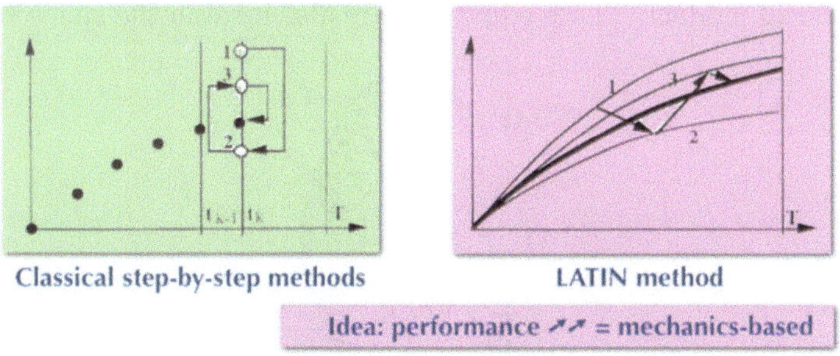

Figure 12. The LATIN method and classical step-by-step methods

What makes the equations different in mechanics? In the process of designing computational strategies, one can face two main undesirable properties:

- the equations can be nonlinear;
- the equations can be global in the space and/or time variables, leading to the resolution of large-scale problems.

We prefer problems which are either linear or local (an equation is local in the space variables if it connects pointwise quantities defined at the same point, e.g. Hooke's law). It is clear that an equation which is local in the space variables leads, after discretization, to small independent systems of equations associated with the Gauss points. This is a very favorable property, in particular regarding parallel computing.

Let us now consider a structural problem defined over the time-space domain $[0, T] \times \Omega$, where the environment of the structure Ω is prescribed through given loads and displacements (figure 16). For the sake of the simplicity, we will assume small displacements and quasi-static conditions. The material model is defined in the classical thermodynamic framework with internal variables. Then, the equations can be divided over the time-space domain $[0, T] \times \Omega$ into:

- equilibrium and compatibility equations;
- constitutive relations (state equations + state evolution laws).

The former are linear equations, while the state evolution laws are local in both the space and time variables; the status of the state equations is less clear. Introducing a change of variables for the internal state variables, we proved that these equations can take a linear form (which we call the "normal" formulation (Ladevèze, 1989; Cognard and Ladevèze, 1993)) for most material models. These equations are also local in the space variables.

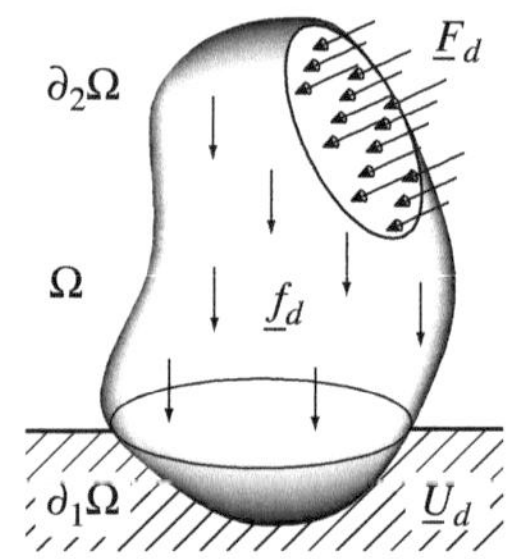

Figure 13. The reference problem and boundary conditions

The LATIN method: principles. The LATIN method operates over the time-space domain $[0, T] \times \Omega$, and its first principle (P1) consists in separating the difficulties. Thus, the equations are divided into:

- a set of linear equations which can be global in the space variables:
  - the equilibrium and compatibility equations;
  - the state equations
- a set of equations which are local in the space variables and can be nonlinear:
  - the state evolution laws.

In the geometric representation of figure 14, $\mathbf{A_d}$ and $\mathbf{\Gamma}$ represent the solutions of the first and second set respectively. The exact solution is the intersection of $\mathbf{\Gamma}$ and $\mathbf{A_d}$. The partitioning into $\mathbf{A_d}$ and $\mathbf{\Gamma}$ is very natural, $\mathbf{A_d}$ being associated with the free energy and $\mathbf{\Gamma}$ with the dissipation.

The second principle (P2) is also very natural. It consists in solving the two sets of equations alternatively until practical convergence. In order to do that, one uses search directions given as parameters. One possible choice (the Newton set) consists of the tangent direction and its conjugate (see figure 15). A major difference compared to classical quasi-Newton

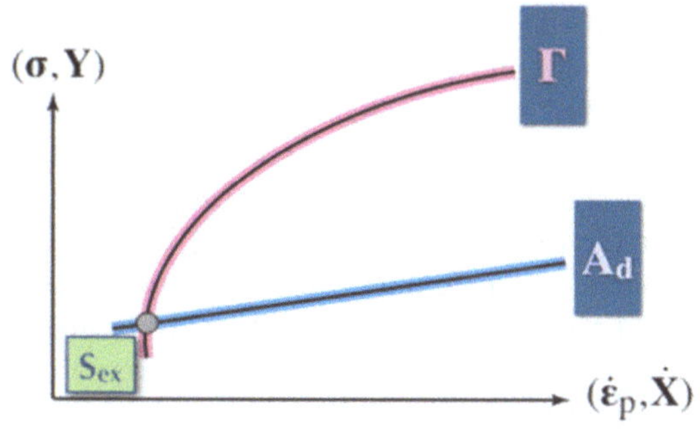

Figure 14. The geometric representation associated with the problem being studied

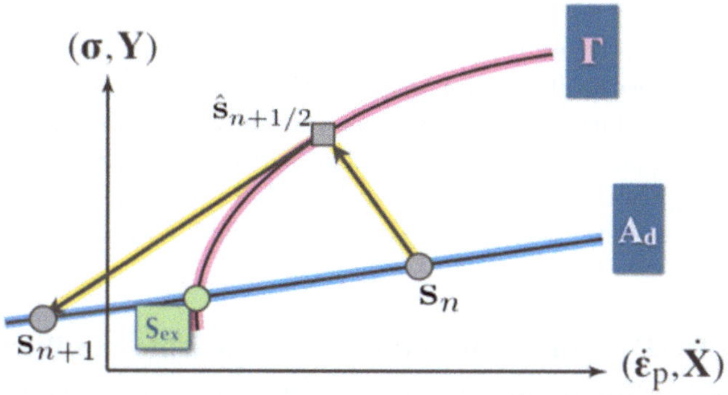

Figure 15. One iteration of the LATIN method over $[0, T] \times \Omega$

techniques is that the tangent does not refer to the stiffness, but to the operator which defines the state evolution laws.

In summary, at the nth iteration, one must solve a linear problem defined over $[0, T] \times \Omega$ in order to get $\mathbf{s}_n$, and a local problem in the space and time variables in order to get $\hat{\mathbf{s}}_{n+1/2}$. The strong convergence of this iteration process has been proven in the case of non-softening materials. The distance between two successive approximate solutions defines a good and easily obtained error indicator.

What makes the LATIN method efficient? The first two principles lead only to a reformulation of the problem being studied, but there is a third principle associated with the resolution of the linear problem defining

$\mathbf{s}_n$, which is a global problem over the time-space domain $[0, T] \times \Omega$. Principle (P3) consists in using what we called "radial loading approximations" (Ladevèze, 1985b, 1989, 1999), in which the solution is described as a finite sum of products of a time function by a space function. F. Chinesta and I have recently coined the name "Proper Generalized Decomposition" (as an extension of Proper Orthogonal Decomposition) for this type of approximation. In this approach, one calculates the solution of the problem and the shape functions used to describe it simultaneously by means of an iterative process. This is what makes the PGD method different. In short, one can say that the shape functions in the PGD method depend on all the data, i.e. both on the operator and on the right-hand side, which is clearly different from POD approximations. Developments of PGD for many other problems could be seen in the last IACM Expression Bulletin (Chinesta, 2009). PGD turns out to be the key to the efficiency of the LATIN method because it drastically reduces both the computation cost and the information storage requirements. Unfortunately, the PGD approximation over the time-space domain works very well only for quasi-static problems. Extensions to large-displacement and large-strain problems were introduced in (Ladevèze, 1999) thanks to an original reformulation which is formally similar to that of small displacement problems. Developments involving complex material models and instabilities have been made (Ladevèze, 1999; Cognard and Ladevèze, 1993).

3.2 The PGD/LATIN method as a monoscale nonlinear solver

The reference problem. In order to illustrate the method, let us consider the case of the quasi-static, isothermal evolution of a structure defined over the time-space domain $I \times \Omega$, where $I = [0, T]$ and $\Omega \subset \mathbb{R}^d$. Under the assumption of small perturbations, the state of the structure is defined by: $\boldsymbol{\varepsilon}_\mathrm{p}$, the inelastic part of the strain field $\boldsymbol{\varepsilon}$ corresponding to the displacement field $\underline{U}$, which uncouples into an elastic part $\boldsymbol{\varepsilon}_\mathrm{e}$ and an inelastic part $\boldsymbol{\varepsilon}_\mathrm{p} = \boldsymbol{\varepsilon} - \boldsymbol{\varepsilon}_\mathrm{e}$; $\mathbf{X}$, the remaining internal variables; $\boldsymbol{\sigma}$, the Cauchy stress field; and $\mathbf{Y}$, the set of variables conjugate of $\mathbf{X}$. All these quantities are defined over the time-space domain $I \times \Omega$ and are assumed to be sufficiently regular. Introducing the following notations for the primal and dual fields:

$$\mathbf{u} = \begin{bmatrix} \underline{U} \\ 0 \end{bmatrix}, \; \mathbf{e}_\mathrm{p} = \begin{bmatrix} \boldsymbol{\varepsilon}_\mathrm{p} \\ -\mathbf{X} \end{bmatrix}, \; \mathbf{e} = \begin{bmatrix} \boldsymbol{\varepsilon} \\ 0 \end{bmatrix}, \; \mathbf{e}_\mathrm{e} = \begin{bmatrix} \boldsymbol{\varepsilon}_\mathrm{e} \\ \mathbf{X} \end{bmatrix}, \text{ so that } \mathbf{e}_\mathrm{p} = \mathbf{e} - \mathbf{e}_\mathrm{e}, \mathbf{f} = \begin{bmatrix} \boldsymbol{\sigma} \\ \mathbf{Y} \end{bmatrix} \tag{40}$$

the mechanical dissipation rate for the entire structure Ω is:

$$\int_\Omega (\dot{\boldsymbol{\varepsilon}}_\mathrm{p} : \boldsymbol{\sigma} - \dot{\mathbf{X}} \cdot \mathbf{Y}) d\Omega = \int_\Omega (\dot{\mathbf{e}}_\mathrm{p} \circ \mathbf{f}) d\Omega \tag{41}$$

where $\cdot$ denotes the contraction associated with the tensorial nature of $\mathbf{X}$ and $\mathbf{Y}$, and $\circ$ denotes the corresponding operator. Let us introduce the bilinear "dissipation" form:

$$\langle \mathbf{s}, \mathbf{s}' \rangle = \int_{I\times\Omega} (1 - \frac{t}{T})(\dot{\mathbf{e}}_{\mathrm{p}} \circ \mathbf{f}' + \dot{\mathbf{e}}'_{\mathrm{p}} \circ \mathbf{f}) \, d\Omega dt \tag{42}$$

along with $\mathbf{E}$ and $\mathbf{F}$, the spaces of the fields $\dot{\mathbf{e}}_{\mathrm{p}}$ and $\mathbf{f}$ which are compatible with (42) (using $\dot{\square}$ to designate the time derivative). These spaces enable us to define $\mathbf{S} = \mathbf{E} \times \mathbf{F}$, the space in which the state $\mathbf{s} = (\dot{\mathbf{e}}_{\mathrm{p}}, \mathbf{f})$ of the structure is being sought.

A normal formulation with internal state variables is used to represent the behavior of the material. The *state law* is assumed to lead to:

$$\mathbf{f} = \mathbf{A}\mathbf{e}_{\mathrm{e}} \quad \text{with} \quad \mathbf{A} = \begin{bmatrix} \mathbf{K} & 0 \\ 0 & \mathbf{\Lambda} \end{bmatrix} \tag{43}$$

where Hooke's tensor $\mathbf{K}$ and the constant, symmetric and positive definite tensor $\mathbf{\Lambda}$ are characteristics of the material and, therefore, operator $\mathbf{A}$ is constant, symmetric and positive definite. The *evolution law*, which can be nonlinear, is assumed to be given by the positive differential operator $\mathbf{B}$ such that:

$$\dot{\mathbf{e}}_{\mathrm{p}} = \mathbf{B}(\mathbf{f}) \tag{44}$$

Let us note that such a behavior formulation is available for most material models. For a detailed description of operators $\mathbf{A}$ and $\mathbf{B}$ in the cases of viscoelastic or viscoplastic materials, the reader can refer to (Ladevèze, 1999).

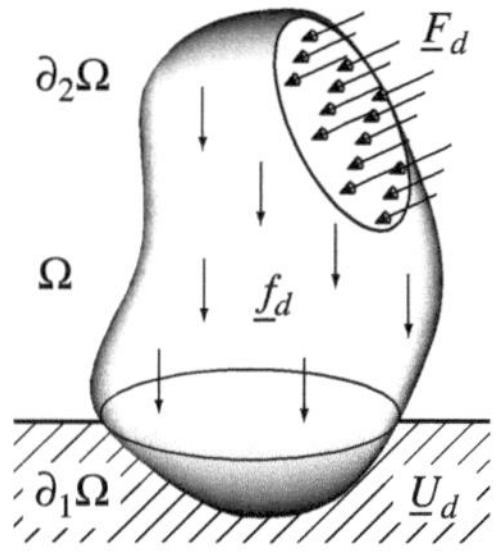

Figure 16. The reference problem and boundary conditions

In this reference problem, the structure is subjected to prescribed body forces $\underline{f}_d$, traction forces $\underline{F}_d$ over a part $\partial_2\Omega$ of the boundary, and displacements $\underline{U}_d$ over the complementary part $\partial_1\Omega$ (see Figure 16). For the sake of simplicity, the displacement $\underline{U}$ alone is assumed to have a nonzero initial value denoted $\underline{U}_0$. In order to formulate the reference problem, let us introduce the following functional subspaces of $\mathbf{S}$ (where $\square^\star$ denotes vector spaces associated with affine spaces):

- **the space $\boldsymbol{\mathcal{U}}$ of the kinematically admissible fields $\mathbf{u}$** whose displacement $\underline{U}$ is equal to the prescribed displacement $\underline{U}_d$ over $\partial_1\Omega$ and verifies the initial conditions $\underline{U}_0$: $\underline{U}_{|\partial_1\Omega} = \underline{U}_d$ and $\underline{U}_{|t=0} = \underline{U}_0$.

- **the space $\boldsymbol{\mathcal{F}}$ of the statically admissible fields $\mathbf{f}$** whose stress field $\boldsymbol{\sigma}$ is in equilibrium with the prescribed external forces $\underline{F}_d$ over $\partial_2\Omega$ and verifies the momentum conservation equation: $\forall \mathbf{u}^\star \in \boldsymbol{\mathcal{U}}^\star$,

$$-\int_{I\times\Omega} \mathbf{f} \circ \mathbf{e}(\underline{\dot{U}}^\star)\, d\Omega dt + \int_{I\times\Omega} \underline{f}_d \cdot \underline{\dot{U}}^\star\, d\Omega dt + \int_{I\times\partial_2\Omega} \underline{F}_d \cdot \underline{\dot{U}}^\star\, dSdt = 0 \tag{45}$$

- **the space $\boldsymbol{\mathcal{E}}$ of the kinematically admissible fields $\dot{\mathbf{e}}$** whose strain field $\boldsymbol{\varepsilon}$ derives from a displacement field $\underline{U}$ belonging to $\boldsymbol{\mathcal{U}}$:

$$\forall \mathbf{f}^\star \in \boldsymbol{\mathcal{F}}^\star,\ -\int_{I\times\Omega} \mathbf{f}^\star \circ \dot{\mathbf{e}}\, d\Omega dt + \int_{I\times\partial_1\Omega} \boldsymbol{\sigma}^\star \underline{n} \cdot \underline{\dot{U}}_d\, dSdt = 0 \tag{46}$$

- **the space $\mathbf{A_d}$ of the admissible fields $\mathbf{s}$** in which $\mathbf{f}$ is statically admissible, $\mathbf{f}$ and $\dot{\mathbf{e}}_\mathrm{p}$ verify the state law (43), and the corresponding $\dot{\mathbf{e}}$ is kinematically admissible:

$$\mathbf{f} \in \boldsymbol{\mathcal{F}},\ (\mathbf{A}^{-1}\dot{\mathbf{f}} + \dot{\mathbf{e}}_\mathrm{p}) \in \boldsymbol{\mathcal{E}} \tag{47}$$

- **the space $\boldsymbol{\Gamma}$ of the fields $\mathbf{s}$** in which $\mathbf{f}$ and $\dot{\mathbf{e}}_\mathrm{p}$ verify the evolution law (44):

$$\dot{\mathbf{e}}_\mathrm{p} = \mathbf{B}(\mathbf{f}) \tag{48}$$

Obviously, the solution $\mathbf{s_{ref}}$ of the problem over the time-space domain $I \times \Omega$ can be viewed as the intersection of $\mathbf{A_d}$ and $\boldsymbol{\Gamma}$. Then, the reference problem becomes:

$$\text{Find } \mathbf{s_{ref}} \in \mathbf{A_d} \cap \boldsymbol{\Gamma} \tag{49}$$

It is important to note that $\mathbf{A_d}$ is a set of solutions of global linear equations, while $\boldsymbol{\Gamma}$ is a set of solutions of (possibly) nonlinear equations which are local in time and in space.

Nearly all the methods available in mechanics for the resolution of nonlinear evolution problems such as (49) are incremental methods. The time interval $I = [0, T]$ being considered is decomposed into a series of small intervals. Assuming that the history of the different quantities until the current time t is known, the objective is to study the new interval $[t, t + \Delta t]$, where

Δt denotes the increment. For example, the assumption that the evolution over $[t, t+\Delta t]$ is linear, *i.e.* that the history depends only on the values at time $t+\Delta t$, leads to a nonlinear problem which does not involve time and which can be solved by a Newton-type method. Roughly speaking, this type of method requires the resolution of about $n_I \times n_N$ spatial problems for the Newton scheme to converge (n_I being the number of time intervals in I and n_N the number of subiterations, which is assumed to be constant).

The LATIN method, which will be described next, relies on a different scheme which takes advantage of the favorable properties of the equations (linearity of $\mathbf{A_d}$ and locality of $\mathbf{\Gamma}$) while avoiding the simultaneous treatment of their difficult aspects (the globality of $\mathbf{A_d}$ and the nonlinearity of $\mathbf{\Gamma}$).

The LATIN method as a solver. Let us apply the LATIN method to the formulation of the problem to be solved presented previously (49). The LATIN method is a general, mechanics-based computational strategy for the resolution of time-dependent nonlinear problems which operates over the entire time-space domain (Ladevèze, 1989, 1999). The solution of the problem is obtained through an iterative scheme. An iteration consists of two stages, called the "local stage" and the "linear stage". As shown in Figure 17, these stages consist in building fields of $\mathbf{\Gamma}$ and $\mathbf{A_d}$ alternatively. Under conditions which will described later, this iterative process converges toward the solution $\mathbf{s_{ref}}$ of the problem. The two stages of the method will be described in detail in the following sections.

$$\cdots \longrightarrow \underbrace{\mathbf{s}_n \in \mathbf{A_d} \xrightarrow{\text{local}} \hat{\mathbf{s}}_{n+1/2} \in \mathbf{\Gamma} \xrightarrow{\text{linear}} \mathbf{s}_{n+1} \in \mathbf{A_d}}_{\text{Iteration } n+1} \longrightarrow \hat{\mathbf{s}}_{n+3/2} \longrightarrow \cdots$$

Figure 17. The local stage and the linear stage of the LATIN method at Iteration $n+1$

It is important to note that at any given iteration $n+1$ of the method two approximate solutions ($\hat{\mathbf{s}}_{n+1/2} \in \mathbf{\Gamma}$ and $\mathbf{s}_{n+1} \in \mathbf{A_d}$) are known over the entire time-space domain $I \times \Omega$. In that sense, the LATIN method is considered to be "nonincremental", contrary to incremental methods, in which an approximate solution is achieved only once the algorithm has been carried out through the entire time interval. Figure 18 illustrates the LATIN algorithm by showing, in the space $\mathbf{S} = \mathbf{E} \times \mathbf{F}$ in which the solution is sought, the linear space $\mathbf{A_d}$ and the nonlinear space $\mathbf{\Gamma}$ along with the solution $\mathbf{s_{ref}}$. It is clear that in order to complete our coverage of the problem we also need to introduce what we call the "search directions" $\mathbf{E}^+$ and $\mathbf{E}^-$. This will be done later.

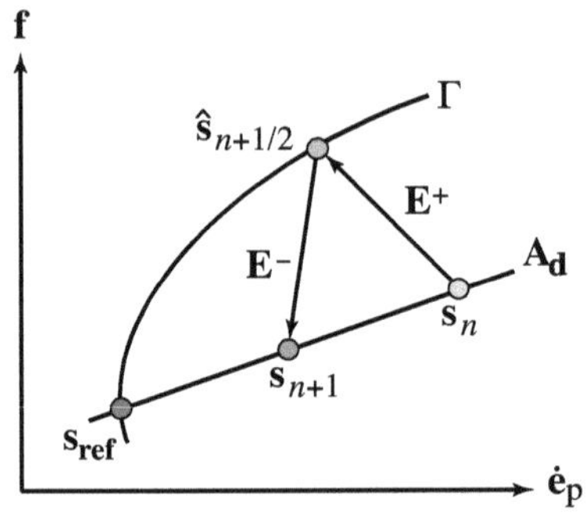

Figure 18. An iteration of the LATIN method

The local stage at Iteration $n + 1$

This stage consists, given $\mathbf{s}_n \in \mathbf{A_d}$, in building $\hat{\mathbf{s}}_{n+1/2} \in \mathbf{\Gamma}$, then using an "ascent" search direction $\mathbf{E}^+$ followed by $\hat{\mathbf{s}}_{n+1/2} - \mathbf{s}_n = \delta\mathbf{s}$ (see Figure 18). This search direction is defined by:

$$\mathbf{E}^+ = \left\{ \ \delta\mathbf{s} = (\delta\dot{\mathbf{e}}_\mathrm{p}, \delta\mathbf{f}) \ \middle| \ \delta\dot{\mathbf{e}}_\mathrm{p} + \mathbf{H}^+\delta\mathbf{f} = 0 \ \right\} \tag{50}$$

where $\mathbf{H}^+$ is a symmetric, positive definite operator which is a parameter of the method. One can easily show that seeking $\hat{\mathbf{s}}_{n+1/2} = (\hat{\dot{\mathbf{e}}}_\mathrm{p}, \hat{\mathbf{f}})$ common to $\mathbf{\Gamma}$ and $\mathbf{E}^+$ leads to the resolution of a set of nonlinear problems:

$$\mathbf{B}(\hat{\mathbf{f}}) + \mathbf{H}^+\hat{\mathbf{f}} = \mathbf{a}_n \tag{51}$$

whose right-hand sides $\mathbf{a}_n = \dot{\mathbf{e}}_{\mathrm{p}n} + \mathbf{H}^+\mathbf{f}_n$ are known at this stage, and which are *local in the space variable.* Therefore, these problems lend themselves to the highest degree of parallelism. This property justifies the term "local" used to designate this stage. Once problems (51) have been solved, one determines $\hat{\dot{\mathbf{e}}}_\mathrm{p}$ using search direction $\mathbf{E}^+$, *i.e.* by setting $\hat{\dot{\mathbf{e}}}_\mathrm{p} = \mathbf{a}_n - \mathbf{H}^+\mathbf{f}$.

The linear stage at Iteration $n + 1$

This stage consists, given $\hat{\mathbf{s}}_{n+1/2} \in \mathbf{\Gamma}$, in building $\mathbf{s}_{n+1} \in \mathbf{A_d}$, then using a "descent" search direction $\mathbf{E}^-$ followed by $\mathbf{s}_{n+1} - \hat{\mathbf{s}}_{n+1/2} = \delta\mathbf{s}$ (see

Figure 18). This search direction is defined by:

$$\mathbf{E}^- = \left\{ \ \delta \mathbf{s} = (\delta \dot{\mathbf{e}}_\mathrm{p}, \delta \mathbf{f}) \ \middle| \ \delta \dot{\mathbf{e}}_\mathrm{p} - \mathbf{H}^- \delta \mathbf{f} = 0 \ \right\} \tag{52}$$

where $\mathbf{H}^-$ is a symmetric, positive definite operator which is a parameter of the method. One can show that seeking $\mathbf{s}_{n+1} = (\dot{\mathbf{e}}_\mathrm{p}, \mathbf{f})$ common to $\mathbf{A_d}$ and $\mathbf{E}^-$ leads to the resolution of a global problem in space: $\forall \mathbf{f}^\star \in \boldsymbol{\mathcal{F}}^\star$,

$$\int_{I\times\Omega} \mathbf{f}^\star \circ (\mathbf{A}^{-1}\dot{\mathbf{f}} + \mathbf{H}^- \mathbf{f})\, d\Omega dt = \int_{I\times\Omega} \mathbf{f}^\star \circ \hat{\mathbf{a}}\, d\Omega dt + \int_{I\times\partial_1\Omega} \boldsymbol{\sigma}^\star \underline{n} \cdot \underline{\dot{U}}_d \, dS dt = 0 \tag{53}$$

whose right-hand side $\hat{\mathbf{a}} = \hat{\dot{\mathbf{e}}}_\mathrm{p} - \mathbf{H}^- \hat{\mathbf{f}}$ is known at this stage and where $\underline{U}_d$ is a prescribed boundary condition. This problem is *linear* (and, therefore, this stage is called the "linear" stage), but it depends on time. Its treatment will be discussed in Section 3.3. Once problem (53) has been solved, one determines $\dot{\mathbf{e}}_\mathrm{p}$ using search direction $\mathbf{E}^-$, *i.e.* by setting $\dot{\mathbf{e}}_\mathrm{p} = \hat{\mathbf{a}} - \mathbf{H}^- \mathbf{f}$.

Convergence properties and criterion

If $\mathbf{B}$ is monotonic, using the fact that $\mathbf{A}$ is symmetric and positive definite, one can derive the following anti-monotony and monotony properties:

$$\begin{gathered} \forall (\mathbf{s}, \mathbf{s}') \in \mathbf{A}_\mathbf{d}^2,\ \langle \mathbf{s} - \mathbf{s}', \mathbf{s} - \mathbf{s}' \rangle \leqslant 0 \\ \forall (\mathbf{s}, \mathbf{s}') \in \boldsymbol{\Gamma}^2,\ \langle \mathbf{s} - \mathbf{s}', \mathbf{s} - \mathbf{s}' \rangle \geqslant 0 \end{gathered} \tag{54}$$

The choice of parameters $\mathbf{H}^+$ and $\mathbf{H}^-$ influences only the convergence of the algorithm, but does not affect the solution. In practice, these parameters are chosen as $\mathbf{H}^+ = \mathbf{H}^- = \mathbf{H}$, where $\mathbf{H}$ is a symmetric, positive definite operator which can vary during the iterative process. In that case, following the proof given in (Ladevèze, 1999) based on the anti-monotony and monotony properties (54), one can prove that the quantity $\frac{1}{2}(\mathbf{s}_{n+1} + \mathbf{s}_n)$ converges toward $\mathbf{s}_\mathbf{ref}$, the solution of Problem (49). If the behavior is linear, one can choose, for example, $\mathbf{H} = \mathbf{B}$. Other possible choices, especially in the nonlinear case, are discussed in (Ladevèze, 1999).

In order to ensure the convergence of $\mathbf{s}_n$ and, more generally, to ensure convergence for many types of material behavior, a relaxation technique may be required. Renaming the quantity previously denoted $\mathbf{s}_{n+1}$ as $\bar{\mathbf{s}}_{n+1}$, we redefine the approximate solution $\mathbf{s}_{n+1}$ generated by linear stage $n+1$ as:

$$\mathbf{s}_{n+1} = \mu \bar{\mathbf{s}}_{n+1} + (1 - \mu) \mathbf{s}_n \tag{55}$$

where μ is a relaxation parameter usually chosen to be equal to 0.8.

Since the reference solution $\mathbf{s_{ref}}$ is the intersection of $\mathbf{\Gamma}$ and $\mathbf{A_d}$, the distance between $\hat{\mathbf{s}}_{n+1/2}$ and $\mathbf{s}_n$ is a good error indicator for assessing the convergence of the algorithm. The simplest measure of that distance is:

$$\eta = \frac{\|\hat{\mathbf{s}}_{n+1/2} - \mathbf{s}_n\|}{\frac{1}{2}\|\hat{\mathbf{s}}_{n+1/2} + \mathbf{s}_n\|} \tag{56}$$

with:

$$\|\mathbf{s}\|^2 = \frac{1}{2}\int_{I\times\Omega} (\dot{\mathbf{e}}_\mathrm{p} \circ \mathbf{H}^{-1}\dot{\mathbf{e}}_\mathrm{p} + \mathbf{f} \circ \mathbf{H}\mathbf{f})\, d\Omega dt \tag{57}$$

3.3 The LATIN method and Proper Generalized Decomposition

Proper Generalized Decomposition: the principle

Problem (53) is solved using the Proper Generalized Decomposition method, which was first introduced in (Ladevèze, 1985b) under the name "radial approximation" (see also (Ladevèze, 1991, 1999)). In order to do that, the linear stage at Iteration $n+1$ is rewritten as a corrective increment $\Delta\mathbf{s}$ to the previous approximation $\mathbf{s}_n$, so that the quantity being sought is no longer $\mathbf{s}_{n+1} \in \mathbf{A_d}$, but $\Delta\mathbf{s}$ such that $\mathbf{s}_{n+1} = (\dot{\mathbf{e}}_\mathrm{p}, \mathbf{f}) = \mathbf{s}_n + \Delta\mathbf{s}$.

If the initial solution $\mathbf{s}_0$ (e.g. the solution of a linear elastic calculation) belongs to $\mathbf{A_d}$, all the corrective increments $\Delta\mathbf{s} = (\Delta\dot{\mathbf{e}}_\mathrm{p}, \Delta\mathbf{f})$ are sought in $\mathbf{A}^\star_\mathbf{d}$, the space which corresponds to $\mathbf{A_d}$ with homogeneous conditions. With the choice $\mathbf{H}^- = \mathbf{H}$ mentioned previously, where $\mathbf{H}$ is a symmetric, positive definite operator, the search direction $\mathbf{E}^-$ defined by (52) becomes:

$$\Delta\dot{\mathbf{e}}_\mathrm{p} - \mathbf{H}\Delta\mathbf{f} - \Delta\hat{\mathbf{a}} = 0 \tag{58}$$

where $\Delta\hat{\mathbf{a}} = (\hat{\dot{\mathbf{e}}}_\mathrm{p} - \dot{\mathbf{e}}_{\mathrm{p}n}) - \mathbf{H}(\hat{\mathbf{f}} - \mathbf{f}_n)$, which is known at this stage, is rewritten as the minimization of a global constitutive relation error in $\mathbf{A}^\star_\mathbf{d}$:

$$\Delta\mathbf{s} = \mathrm{Arg} \min_{\Delta\mathbf{s}\in\mathbf{A}^\star_\mathbf{d}} e^2_{CR}(\Delta\mathbf{s}) = \|\Delta\dot{\mathbf{e}}_\mathrm{p} - \mathbf{H}\Delta\mathbf{f} - \Delta\hat{\mathbf{a}}\|^2_\mathbf{M} \tag{59}$$

This expression introduces the norm:

$$\|\square\|^2_\mathbf{M} = \int_{I\times\Omega} \square \circ \mathbf{M}\square \, d\Omega dt \tag{60}$$

where $\mathbf{M}$ a symmetric, positive definite operator usually chosen to be equal to $\mathbf{M} = (1 - \frac{t}{T})\mathbf{H}$. The new formulation (59) is equivalent to the initial problem (53) provided no approximation is introduced in seeking the corrective increment $\Delta\mathbf{s} = (\Delta\dot{\mathbf{e}}_\mathrm{p}, \Delta\mathbf{f})$.

input: the previous approximation $\mathbf{e}_{\mathrm{p}n} = a_0\mathbf{E}_{\mathrm{p}0} + \sum_{i=1}^{n} a_i\mathbf{E}_{\mathrm{p}i}$ and the known quantities $\Delta\hat{\mathbf{a}}$

Step 1: use of the reduced basis
keeping the space functions $(\mathbf{E}_{\mathrm{p}i})_{i=1}^{n}$ fixed, one seeks the $(a_i)_{i=1}^{n}$ which minimize e_{CR}^2;

Step 2: addition of new functions
initialization: $a^0(t)$ (for example, $f(t) = \alpha t$);
for $k = 1$ **to** k_{max} **do**

- **spatial problem:** given the time function $a^k(t)$, one seeks the $\mathbf{E}_{\mathrm{p}}^k$ which minimize e_{CR}^2;
- **time problem:** given the space functions $\mathbf{E}_{\mathrm{p}}^k$, one seeks $a^k(t)$ which minimizes e_{CR}^2;
- **orthonormalization** of the space functions with respect to the previous spatial basis;

end

Algorithm 1: Iterative generation of the PGD corrective terms at Iteration $n + 1$

Step 1: use of the reduced basis

This POD phase consists in formulating an inexpensive prediction thanks to one's knowledge of the solutions from previous iterations. This approximate solution is introduced into the constitutive relation error (59), but in this case the only unknowns are the time functions. In other words, one seeks the optimum combination of the reduced basis of space functions which minimizes the constitutive relation error e_{CR}^2. These time functions verify a system of linear differential equations in time with conditions at $t = 0$ and $t = T$, whose solution is obtained classically. This is usually a relatively small system.

Step 2: addition of new functions

The initialization of the iterative process is done considering as a time function the standard error. The minimization with respect to the space variables leads to the resolution of a time-independent spatial problem defined over Ω. That is a classical FE problem. The minimization with respect to the time variable leads to a scalar differential equation defined over I, with conditions at the two extremities of the time interval, whose resolution is quite inexpensive. An easier way for not many time steps is to solve the global problem derived from the minimization. The iterative process is followed until few iterations, practically 2. Once these new space and time functions have been calculated, the space functions are orthogonalized and added to the reduced basis.

Since the construction of the space functions is by far the most expensive part of the process, it is advisable to store and reuse these functions. Thus, the space functions constructed up to Iteration n can be reused at Iteration $n+1$ and Step 1 is executed systematically only to update the corresponding time functions. Conversely, Step 2, which generates new time and space functions, can be skipped if the correction from Step 1 is sufficient. (This technique requires an *ad hoc* criterion to be set up.)

Discussion of the computational cost

The resolution of Problem (53) using a standard incremental technique would have required the resolution of n_I spatial problems at each iteration of the LATIN algorithm, n_I being the number of time intervals in I. Then, the generation of the final approximation of the solution $\mathbf{s_{ref}}$ would have required the resolution of $n_I \times n_{iter}$ spatial problems, n_{iter} being the number of iterations of the LATIN method necessary for a given accuracy assessed by the convergence of indicator η. The latter would have been by far the costliest part of the algorithm. In that form, the LATIN method could be viewed as a Newton-type algorithm except for an inversion of the order in which the iterations corresponding to the treatment of the nonlinearity and those corresponding to the treatment of the time evolution are performed, which would result in no decrease in computation costs.

When the PGD technique is used, as can be seen in our examples, Algorithm 1 converges very quickly. Therefore, in practice, k_{max} is set to be equal to only 2 or 3. Then, the generation of the final approximation of the solution $\mathbf{s_{ref}}$ requires the resolution of $k_{max} \times n_{iter}$ spatial problems, and n_{iter} (the number of iterations of the LATIN method necessary for a given accuracy assessed by the convergence of indicator η) is nearly the same as in the case of an incremental strategy. Thus, the decrease in computation cost compared to a Newton-type algorithm is obvious.

3.4 An engineering application

We consider a true engineering example to illustrate the performances of the strategy presented (Relun et al., 2013). This is a relatively small case with 151,600 DOFs and 60 time steps. The geometry here is freely inspired from the true one, a blade of the Vulcain engine.

The blade test-case. This first test-case is freely inspired from a blade of the Vulcain engine. The geometry, boundary conditions and mesh are presented on figure 19. The typical dimensions are 45 (x direction) $\times$ 70 (y direction) $\times$ 50 (z direction) mm. A two sinusoidal cycles displacement with an amplitude of 0.3 mm ($\pm$0.15 mm) is prescribed. The total number of dofs is about 141,500 and the sinusoidal cycle is discretized by 60 time steps. The material coefficients used for the Marquis-Chaboche elastic-viscoplastic material (with kinematic hardening and Norton power law) are the following: $E = 191,500\ MPa$; $\nu = 0,3$; $R_0 = 25\ MPa$; $C = 131,000$; $\gamma = 310$; $K = 600\ MPa.s^{(1/N)}$; $N = 10$. These values are typical of a Titanium TA 6 V material at 500 K.

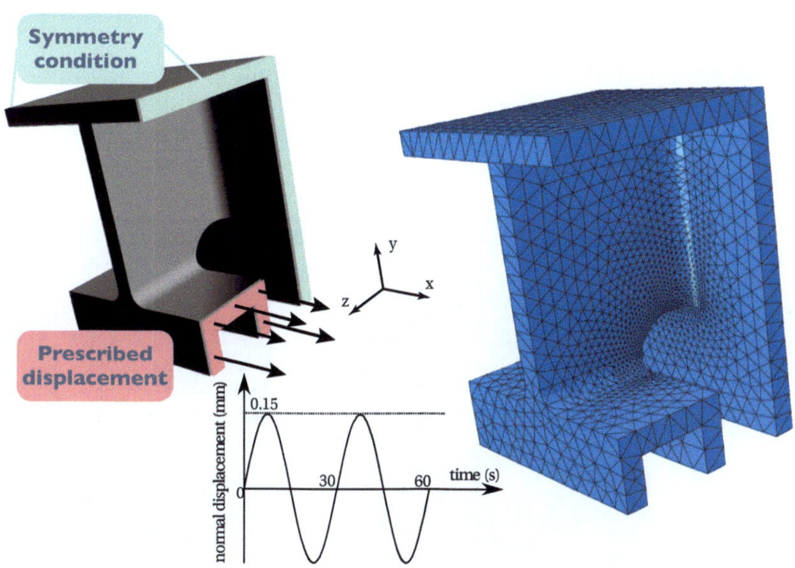

Figure 19. Geometry, boundary conditions and mesh of the blade test-case

The problem is not only strongly nonlinear and time-dependent. There also parameters: γ, the yield stress R_0 and the loading amplitude. They values are not very well known. The range of variation of each parameter was discretized into 10 arbitrarily values, leading to 1,000 different nonlinear time-dependent problems (see table 1).

Parameter	Min. value	Max. value	step	variation
$\pm$ loading amplitude	0.1 mm	0.19 mm	0.01 mm	$\pm$31 %
yield stress R_0	20 MPa	29 MPa	1 MPa	$\pm$18 %
γ	285	330	5	$\pm$7 %

Table 1. Range of variation of loading amplitude, R_0 and γ

Comparison PGD/LATIN and ABAQUS. A first run is carried out and we compare on figure 20 the Von Mises fields obtained with ABAQUS and the PGD/LATIN method with 17 PGD modes. The comparison shows that there is no appreciable difference between this two fields (the same scale is used for the two plots). A quantitative comparison is presented on figure 21: the stress versus strain curve is plotted at the most loaded gauss point. Again, the two curves obtained with ABAQUS and the LATIN method are almost identical.

These results enable the validity of the two models to be verified. Moreover, the computations have been carried out on an Intel bi-Xeon processor (total of 12 cores) at 2.8 GHz with 12 GB of RAM. On this single run, CPU time with ABAQUS was 2200 s and CPU time with the LATIN method software was 1600 s: more than 1.35 time faster than ABAQUS.

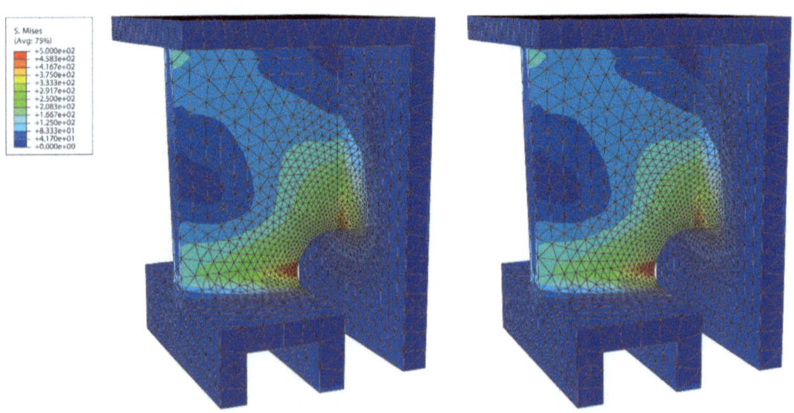

Figure 20. Comparison of the Von Mises fields obtained at time step 55 with the PGD/LATIN (left) and ABAQUS (right).

Virtual Chart. Figure 22 gives the virtual chart related to the maximum value of the Von Mises stress. The exploration of the design space is ordered:

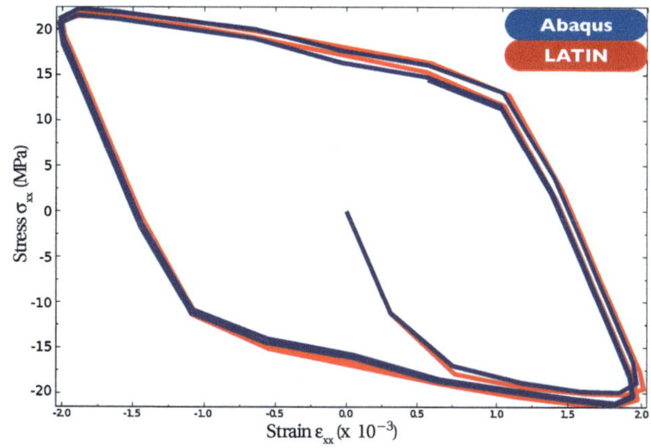

Figure 21. Stress vs strain curves at the most loaded gauss point

only one parameter changes from one set to another. The evolutions of the values of the parameters, versus the number of the set of parameters are presented on figure 23.

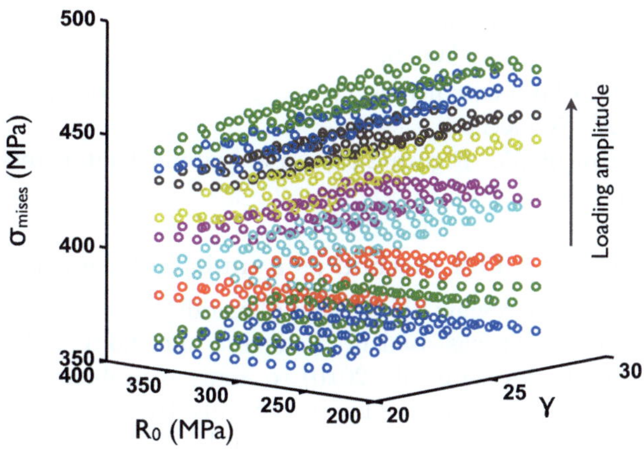

Figure 22. Virtual chart giving the maximum Von Mises stress as a function of the parameters

The technique used is very simple. For a new set of parameters, one starts with the solution obtained for the previous set as initialization. That is possible thanks to the PGD/LATIN method. The performances of the

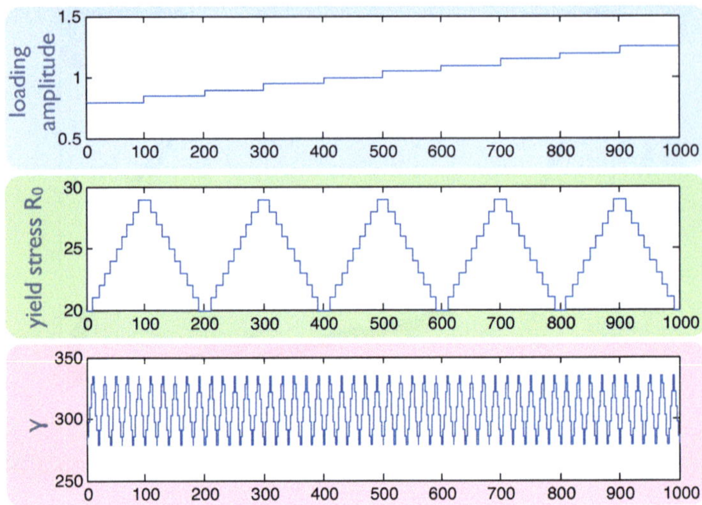

Figure 23. Evolutions of the set of parameters vs set numbers

multiple runs algorithm are the following:

- about 25 days (estimated time) are necessary to complete the 1,000 resolutions with ABAQUS;
- less than 17 hours are necessary to complete the 1,000 resolutions with PGD/LATIN method.

It follows a gain of 35. Figure 24 gives a graphical interpretation of the strategy used.

3.5 The PGD/LATIN method as a multiscale nonlinear solver

In this section, we review how Proper Generalized Decomposition within the LATIN framework was used in (Ladevèze and Nouy, 2003; Nouy and Ladevèze, 2004; Ladevèze et al., 2010) to analyze problems involving two or more very different scales. This is typically the case when domain Ω in Figure 16 is highly heterogeneous and the local solution involves short-wavelength phenomena in both space and time. Attempting to solve such problems with classical finite element codes leads to systems with very large numbers of degrees of freedom, and the associated computation costs are generally prohibitive. Therefore, one of today's main challenges is to derive computational strategies capable of solving such engineering problems through true interaction between the microscale and the macroscale in both space and time.

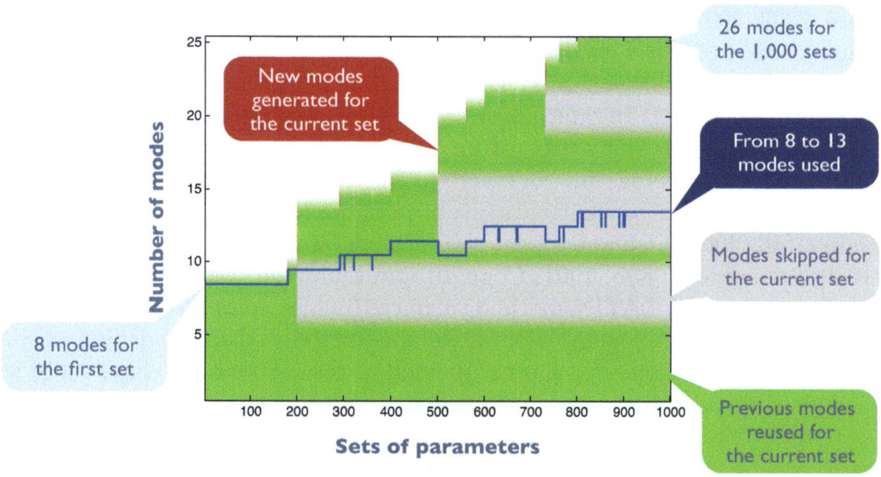

Figure 24. Graphical interpretation of the strategy used for multiple runs

The key issue is the transfer of information from one scale to the other. A very efficient strategy for linear periodic media consists in applying the homogenization theory introduced by Sanchez-Palencia (Sanchez-Palencia, 1974, 1980), for which additional developments and related computational approaches can be found in (Feyel, 2003; Devries et al., 1989; Zohdi et al., 1996; Oden et al., 1999; Fish et al., 1997; Lefik and Schrefler, 2000). First, the resolution of the macro problem leads to effective values of the unknowns; then, the micro solution is calculated locally based on the macro solution. These approaches are often named FE^2. The fundamental assumption, besides periodicity, is that the ratio of the characteristic length of the microscale to that of the macroscale is small. Boundary zones, in which the material cannot be homogenized, require special treatment. Moreover, this theory cannot be applied directly to time-dependent nonlinear problems. Other computational strategies using homogenization techniques based on the Hill-Mandel conditions (Huet, 1990) have also been proposed (Kouznetsova et al., 2002; Zohdi and Wriggers, 2005), but have similar limitations. Other paradigms for building multiscale computational strategies can be found in (Hughes, 1995; Ladevèze et al., 2001). All these approaches introduce different scales in space alone.

Comparatively fewer works have been devoted to multi-time-scale computational strategies. What one calls multi-time-step methods (Combescure and Gravouil, 2002; Faucher and Combescure, 2003; Gravouil and Combescure, 2003, 2001) and time-decomposed parallel time integrators (Farhat and Chandesris, 2003; Belytschko et al., 1985) deal with different time discretizations and integration schemes. Local enrichment functions were introduced in (Bottasso, 2002). In multiphysics problems, a coupling among time grids may be envisaged. This type of problem was solved in (Dureisseix et al., 2003a) by introducing "micro/macro projectors" among grids. Some works have been devoted to the treatment of periodic loading histories (Maitournam et al., 2002; Fish and Chen, 2001; Guennouni, 1988; Akel and Nguyen, 1989; Ladevèze, 1985b, 1991; Cognard and Ladevèze, 1993).

Regarding the LATIN method, a micro/macro computational strategy involving space homogenization alone was introduced in (Ladevèze et al., 2001). This technique was later expanded in (Ladevèze and Nouy, 2003) to include time as well as space. The advantages of using PGD in this context were described in (Nouy and Ladevèze, 2004), and a more efficient and robust version of that strategy in the case of material models with internal variables was introduced in (Ladevèze et al., 2010). The following presentation covers only the main points. A comparison with other multiscale strategies can be found in (Ladevèze et al., 2007).

Reformulation of the reference problem. In order to take advantage of parallel computing and derive a multiscale strategy in time, the time domain I is divided into a small number of coarse subintervals $I_i^C = [t_i^C, t_{i+1}^C]$ (see Figure 25). In the space domain, we follow the classical framework of domain decomposition methods and decompose the structure Ω into an assembly of substructures Ω_E[1] and interfaces $\Phi_{EE'}$[2] as shown on Figure 25.

The state of a substructure Ω_E is defined entirely by $\dot{\mathbf{e}}_{\mathrm{p}E}$ and $\mathbf{f}_E$, the restriction of fields $\dot{\mathbf{e}}_{\mathrm{p}}$ and $\mathbf{f}$ to Ω_E. The state of an interface $\Phi_{EE'}$ is defined by its own variables — the velocity distributions $(\underline{\dot{W}}_E, \underline{\dot{W}}_{E'})$ and force distributions $(\underline{F}_E, \underline{F}_{E'})$ on both sides of the interface — and its own behavior. This behavior consists of the equilibrium condition (the action-reaction principle) plus a constitutive relation, defined by operator $\mathbf{b}_{EE'}$, which characterizes the behavior of the interface. In other words, the pairs of fields $(\underline{\dot{W}}_E, \underline{\dot{W}}_{E'})$ and $(\underline{F}_E, \underline{F}_{E'})$ must verify:

$$\begin{aligned} &\underline{F}_E + \underline{F}_{E'} = \underline{0} \\ &\mathbf{b}_{EE'}(\underline{\dot{W}}_E, \underline{\dot{W}}_{E'}, \underline{F}_E, \underline{F}_{E'}) = 0 \end{aligned} \tag{61}$$

In the case of a perfect connection, (61) is very simple and can be ex-

[1]the notation $\square_E$ designates the restriction of a quantity $\square$ to substructure Ω.

[2]$\Phi_{EE'}$ designates the interface between two substructures Ω_E and $\Omega_{E'}$.

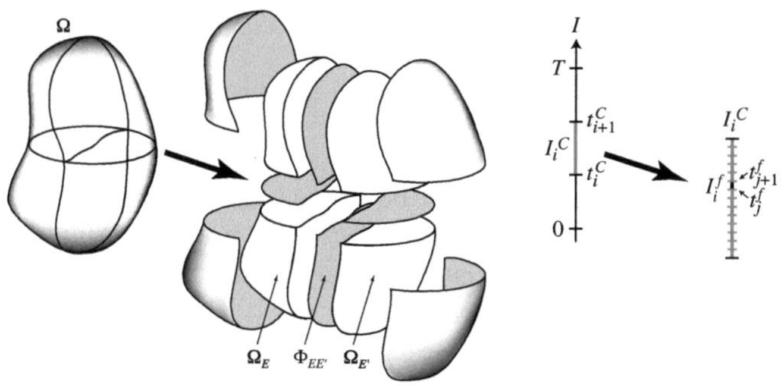

Figure 25. Decomposition of Ω into substructures Ω_E and interfaces $\Phi_{EE'}$, and decomposition of I into subintervals I_i^C

pressed as follows:

$$\begin{aligned} \underline{F}_E + \underline{F}_{E'} &= \underline{0} \\ \underline{\dot{W}}_E - \underline{\dot{W}}_{E'} &= \underline{0} \end{aligned} \tag{62}$$

In the case of unilateral contact with or without friction, the reader can refer to (Dureisseix and Farhat, 2001) for a detailed expression of operator $\mathbf{b}_{EE'}$. The interface concept can also be extended to the boundary of Ω over which a surface displacement distribution or a surface force distribution is prescribed. In order to do that, one simply takes $\underline{W}_E = \underline{U}_d$ for a displacement prescribed over $\Phi_{E1} = \partial\Omega_E \cap \partial_1\Omega$, and $\underline{F}_E = \underline{F}_d$ for a force prescribed over $\Phi_{E2} = \partial\Omega_E \cap \partial_2\Omega$.

Let $\mathbf{s}_E = (\dot{\mathbf{e}}_{\mathrm{p}E}, \underline{\dot{W}}_E, \mathbf{f}_E, \underline{F}_E)$ denote the set of the fields which describe the state of substructure Ω_E and its boundary $\partial\Omega_E$, and let $\mathbf{E}_E$, $\boldsymbol{\mathcal{W}}_E$, $\mathbf{F}_E$ and $\boldsymbol{\mathcal{F}}_E$ denote the corresponding spaces. The problem consists in finding $\mathbf{s} = (\mathbf{s}_E)_{\Omega_E \subset \Omega}$.

A two-scale description of the unknowns

Two scales, denoted the "macro" scale ($\square^M$) and the "micro" scale ($\square^m$) are introduced in order to describe the interface unknowns. This approach is physically consistent: the macro quantities are mean values over space and time, and due to Saint Venant's principle the effects of the micro complements are localized in space. $\forall \underline{F}^{M\star} \in \boldsymbol{\mathcal{F}}_E^M$, the macro part $\underline{\dot{W}}_E^M$ and the micro complement $\underline{\dot{W}}_E^m$ of $\underline{\dot{W}}_E \in \boldsymbol{\mathcal{W}}_E$ are defined by:

$$\int_{I_i^C \times \partial\Omega_E} (\underline{\dot{W}}_E^M - \underline{\dot{W}}_E) \cdot \underline{F}^{M\star}\, dSdt = 0 \quad \text{and} \quad \underline{\dot{W}}_E^m = \underline{\dot{W}}_E - \underline{\dot{W}}_E^M \tag{63}$$

where the spaces $\boldsymbol{\mathcal{F}}_E^M$ and $\boldsymbol{\mathcal{W}}_E^M$ can be chosen arbitrarily. In practice, these spaces are defined by the linear part in space and the quadratic part in time of the forces and displacements (see Figure 26 for the shape of the corresponding basis functions).

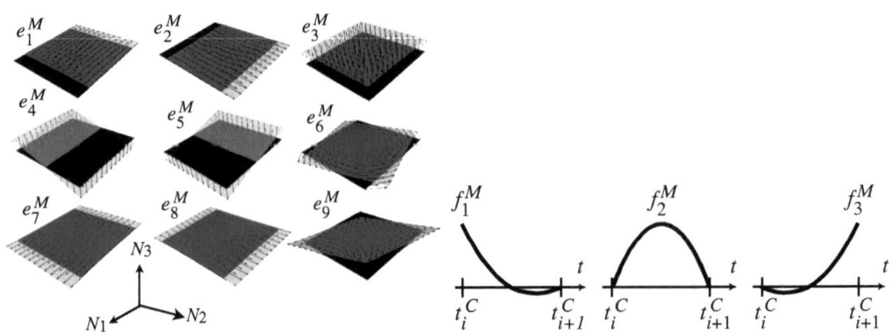

Figure 26. The linear macro basis in space for a square interface $\Phi_{EE'}$ in the 3D case, and the quadratic macro basis in time over I_i^C

An important point of the strategy which leads to its multiscale character is the choice of the admissibility conditions for the macro quantities. The set of the macro forces $(\underline{F}_E^M)_{\Omega_E \subset \Omega}$ is required to verify *a priori* the transmission conditions, including the boundary conditions, systematically:

$$\begin{aligned} \underline{F}_E^M + \underline{F}_{E'}^M &= \underline{0} \quad \text{over } \Phi_{EE'} \\ \underline{F}_E^M - \underline{F}_d^M &= \underline{0} \quad \text{over } \Phi_{E2} = \partial_E\Omega \cap \partial_2\Omega \end{aligned} \tag{64}$$

The corresponding subspace of $\boldsymbol{\mathcal{F}}^M = \bigotimes \boldsymbol{\mathcal{F}}_E^M$ is designated by $\boldsymbol{\mathcal{F}}_{ad}^M$. We use the same definition for $\boldsymbol{\mathcal{W}}^M$ and $\boldsymbol{\mathcal{W}}_{ad}^M$.

Reformulation of the reference problem

Now let us extend the definition of the spaces $\boldsymbol{\mathcal{U}}$, $\boldsymbol{\mathcal{F}}$, $\boldsymbol{\mathcal{E}}$ and $\mathbf{A_d}$ introduced in Section 3.1 in order to reformulate the problem in the decomposed framework. In order to do that, we introduce the following new functional subspaces (whose definitions are obvious):

- **the space $\boldsymbol{\mathcal{U}}_E$ of the kinematically admissible fields** $(\mathbf{u}_E, \underline{W}_E)$, whose displacements $\underline{W}_E$ over $\partial\Omega_E$ are equal to field $\underline{U}_E$ and which verify the initial conditions $\underline{U}_{E0}$: $\underline{U}_{E|\partial\Omega_E} = \underline{W}_E$ and $\underline{U}_{E|t=0} = \underline{U}_{E0}$

- **the space $\boldsymbol{\mathcal{F}}_E$ of the statically admissible fields** $(\mathbf{f}_E, \underline{F}_E)$, whose stress fields $\boldsymbol{\sigma}_E$ are in equilibrium with the interface forces $\underline{F}_E$ over $\partial\Omega_E$ and which verify the momentum conservation equation:

$$\forall(\mathbf{u}_E^\star, \underline{W}_E^\star) \in \boldsymbol{\mathcal{U}}_E^\star, \; -\int_{I_i^C \times \Omega_E} \mathbf{f}_E \circ \mathbf{e}(\underline{\dot{U}}_E^\star)\, d\Omega dt + \int_{I_i^C \times \Omega_E} \underline{f}_d \cdot \underline{\dot{U}}_E^\star \, d\Omega dt + \int_{I_i^C \times \partial\Omega_E} \underline{F}_E \cdot \underline{\dot{W}}_E^\star \, dSdt = 0 \tag{65}$$

- **the space $\boldsymbol{\mathcal{E}}_E$ of the kinematically admissible fields** $(\dot{\mathbf{e}}_E, \underline{\dot{W}}_E)$, whose strain fields $\boldsymbol{\varepsilon}_E$ derive from displacement fields $(\underline{U}_E, \underline{W}_E)$ belonging to $\boldsymbol{\mathcal{U}}_E$:

$$\forall(\mathbf{f}_E^\star, \underline{F}_E^\star) \in \boldsymbol{\mathcal{F}}^\star, \; -\int_{I_i^C \times \Omega_E} \mathbf{f}_E^\star \circ \dot{\mathbf{e}}_E \, d\Omega dt + \int_{I_i^C \times \partial_1\Omega_E} \underline{F}_E^\star \cdot \underline{\dot{W}}_E \, dSdt = 0 \tag{66}$$

- **the space $\mathbf{A}_{\mathbf{d}E}$ of the E-admissible fields $\mathbf{s}_E$** in which $(\mathbf{f}_E, \underline{F}_E)$ is statically admissible, $\mathbf{f}_E$ and $\dot{\mathbf{e}}_{\mathrm{p}E}$ verify the state law (43), and the corresponding $(\dot{\mathbf{e}}_E, \underline{\dot{W}}_E)$ is kinematically admissible:

$$(\mathbf{f}_E, \underline{F}_E) \in \boldsymbol{\mathcal{F}}_E, \; (\mathbf{A}^{-1}\dot{\mathbf{f}}_E + \dot{\mathbf{e}}_{\mathrm{p}E}, \underline{\dot{W}}_E) \in \boldsymbol{\mathcal{E}}_E \tag{67}$$

Then, we also redefine:

- **the space $\mathbf{A_d}$ of the admissible fields** $\mathbf{s} = (\mathbf{s}_E)_{\Omega_E \subset \Omega}$ in which each $\mathbf{s}_E$ is E-admissible and the macro forces verify the transmission conditions (64):

$$\forall\Omega_E \subset \Omega, \; \mathbf{s}_E \in \mathbf{A}_{\mathbf{d}E} \quad \text{and} \quad (\underline{F}_E^M)_{\Omega_E \subset \Omega} \in \boldsymbol{\mathcal{F}}_{ad}^M \tag{68}$$

- **the space $\boldsymbol{\Gamma}$ of the fields** $\mathbf{s} = (\mathbf{s}_E)_{\Omega_E \subset \Omega}$ in which $\mathbf{f}_E$ and $\dot{\mathbf{e}}_{\mathrm{p}E}$ verify the evolution law (44) and $(\underline{F}_E, \underline{F}_{E'}, \underline{\dot{W}}_E, \underline{\dot{W}}_{E'})$ verify the interface behavior (61):

$$\forall\Omega_E \subset \Omega, \; \dot{\mathbf{e}}_{\mathrm{p}E} = \mathbf{B}(\mathbf{f}_E) \; ; \; \forall\Phi_{EE'}, \begin{cases} \underline{F}_E + \underline{F}_{E'} = \underline{0} \\ \mathbf{b}_{EE'}(\underline{\dot{W}}_E, \underline{\dot{W}}_{E'}, \underline{F}_E, \underline{F}_{E'}) = 0 \end{cases} \tag{69}$$

Obviously, the solution $\mathbf{s_{ref}}$ of the problem over the time-space domain $I \times \Omega$ is the intersection of $\mathbf{A_d}$ and $\boldsymbol{\Gamma}$. Thus, the reference problem becomes:

$$\text{Find } \mathbf{s_{ref}} \in \mathbf{A_d} \cap \boldsymbol{\Gamma} \tag{70}$$

The LATIN method as a solver. The previous problem is solved using the LATIN method as described in Section 3.1. In order to do that, one must introduce in the search directions $\mathbf{E}^+$ and $\mathbf{E}^-$ some additional terms corresponding to the boundary fields:

$$\mathbf{E}^+ = \left\{ \delta\mathbf{s} = (\delta\mathbf{s}_E)_{\Omega_E \subset \Omega} \;\middle|\; \begin{array}{l} \delta\dot{\mathbf{e}}_{\mathrm{p}E} + \mathbf{H}^+ \delta\mathbf{f}_E = 0 \\ \delta\underline{\dot{W}}_E - \mathbf{h}^+ \delta\underline{F}_E = \underline{0} \end{array} \right\} \tag{71}$$

and

$$\mathbf{E}^- = \left\{ \delta\mathbf{s} = (\delta\mathbf{s}_E)_{\Omega_E \subset \Omega} \;\middle|\; \begin{array}{l} \delta\dot{\mathbf{e}}_{\mathrm{p}E} - \mathbf{H}^- \delta\mathbf{f}_E = 0 \\ \delta\underline{\dot{W}}_E + \mathbf{h}^- \delta\underline{F}_E = \underline{0} \end{array} \right\} \tag{72}$$

where $\mathbf{h}^+$ and $\mathbf{h}^-$ are symmetric, positive definite operators which are parameters of the method and which are usually set to be equal to $\mathbf{h}^+ = \mathbf{h}^- = \mathbf{h}$. In the case of linear behavior, one can choose $\mathbf{h} = \frac{L}{ET}\mathbf{I}$, where E is the Young's modulus of the material, L_E a characteristic length of the interfaces, T the duration of the phenomenon being studied and $\mathbf{I}$ the identity operator. Other possible choices, especially in the nonlinear case, are discussed in (Ladevèze, 1999).

Then, the local stage becomes very similar to that described in Section 3.1, but the linear stage requires special treatment. The admissibility of the macro forces $(\underline{F}_E^M)_{\Omega_E \subset \Omega} \in \boldsymbol{\mathcal{F}}_{ad}^M$ (and, consequently, the verification of the transmission conditions (64)) is obtained by introducing a Lagrange multiplier $\underline{\dot{\widetilde{W}}}^M = (\underline{\dot{\widetilde{W}}}_E^M)_{\Omega_E \subset \Omega} \in \boldsymbol{\mathcal{W}}_{ad}^{M\star}$. This Lagrange multiplier is obtained by solving a homogenized linear time-space "macro" problem defined over the whole set of interfaces and the entire coarse subinterval I_i^C. Once this macro field is known, the complete solution $\mathbf{s} = (\mathbf{s}_E)_{\Omega_E \subset \Omega}$ is obtained by solving a set of "micro" problems defined over each time-space domain $I_i^C \times \Omega_E$. The detailed treatment of the linear stage using the PGD framework can be found in (Ladevèze et al., 2010).

An illustrative example

The following example, taken from (Ladevèze et al., 2009), illustrates the different steps of the LATIN multiscale strategy and the effectiveness and efficiency of the PGD technique in the resolution of the micro problems introduced previously. Let us consider the 3D problem of a composite structure with cracks (see Figure 27(a)). This structure is fixed at the base and subjected to forces $\underline{F}_1$, $\underline{F}_3$ and $\underline{F}_3$ (see Figure 27(b)). The overall dimensions are $120 \times 120 \times 20$ mm, and the time interval being studied is $T = 10$ s. The cracks are described using unilateral contact with Coulomb friction characterized by parameter $f = 0.3$.

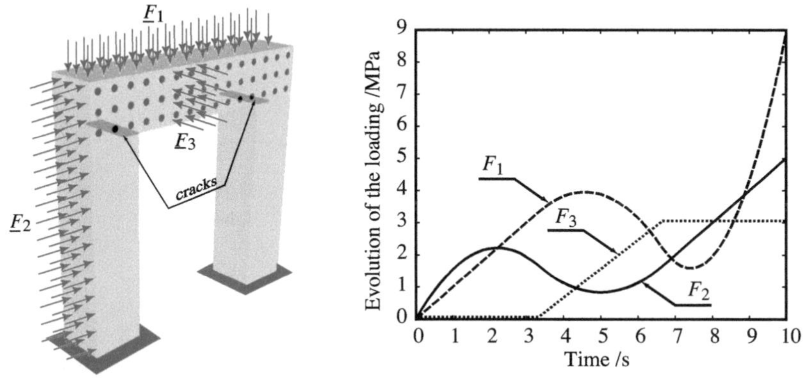

(a) Geometry and loading (b) Loads $\underline{F}_1$, $\underline{F}_3$ and $\underline{F}_3$

Figure 27. Description of the problem

The structure consists of two types of cells: Type-I cells, which are homogeneous and made of a material denoted Type-1, and Type-II cells composed of a matrix of Type-1 material with inclusions of a material denoted Type-2. Both materials are viscoelastic and their properties are given in Table 2. The corresponding constitutive relations are $\dot{\boldsymbol{\varepsilon}}_p = \mathbf{B}_i\boldsymbol{\sigma} = \frac{1}{\eta_i}\mathbf{K}_i^{-1}\boldsymbol{\sigma}$.

Table 2. Material properties

Material	Type-1	Type-2
Young's modulus	$E_1 = 50$ GPa	$E_2 = 250$ GPa
Poisson's ratio	$\nu_1 = 0.3$	$\nu_2 = 0.2$
Viscosity parameter	$\eta_1 = 10$ s	$\eta_2 = 1{,}000$ s

As shown in Figure 28, the problem was divided into 351 substructures (each substructure corresponding to one cell) and 1,296 interfaces. On the micro level, Type-I and Type-II substructures and their interfaces were meshed with 847, 717 and 144 degrees of freedom (DOFs) respectively. In the space domain, the macro part consisted of a single linear element per interface with only 9 DOFs. With respect to time, the micro level was associated with a refined discretization into 60 intervals using a zero-order discontinuous Galerkin scheme, and the macrolevel was associated with a coarse discretization into 3 macro intervals using a second-order discontinuous Galerkin scheme. Because of the linearity of the constitutive relation, the search direction chosen for the substructures was $\mathbf{H} = \mathbf{B}$. With the characteristic length of the interfaces being $L_E = 4$ mm, we chose the search direction $\mathbf{h} = h\mathbf{I}$ for all the interfaces, $h = \frac{L_E}{E_1 \nu_1}$ being a constant scalar.

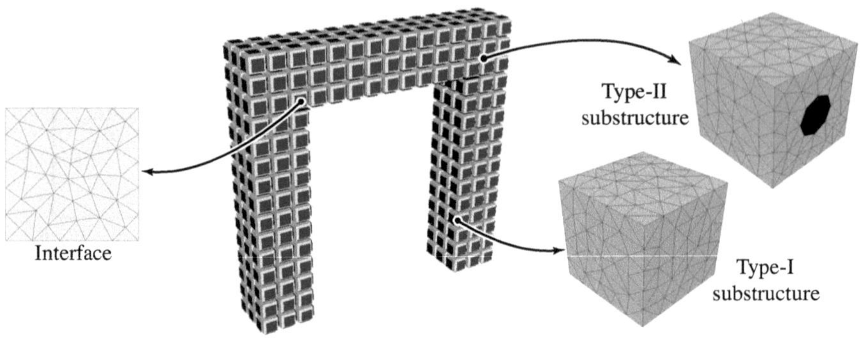

Figure 28. Decomposition into substructures and interfaces and spatial discretization on the microscale

Figure 29 shows the evolution of the error indicator η with the number of iterations. One can observe that the algorithm converges rapidly toward an accurate solution (1% error after 12 iterations). Figure 30 shows the approximate Von Mises stress fields over the structure at the final time $T = 10$ s (with a zoom near one of the cracks) at Iterations 1, 5 and 20 and after convergence (the reference solution) and also shows the evolution over time of the displacement field $\underline{W}$ at point P_2. One can observe that even after the first iteration, thanks to the resolution of the macro problem, the method yields a relatively good approximate solution of the problem in both the space domain and the time domain. After a few iterations, the solution becomes even more accurate and the stress and displacement discrepancies tend toward zero. After 20 iterations, the difference between the approximate solution and the reference solution is no longer visible.

In order to illustrate how the PGD technique was used in the multiscale strategy, let us examine the treatment of one of the micro problems in detail. The micro problem being considered is defined over $I_i^C \times \Omega_E$, where Ω_E corresponds to one of the Type-II substructures of Figure 28. The loading is represented by the distribution of the Lagrange multiplier $\widetilde{\underline{W}}_E^M$ alone, which is not known but depends on few scalar parameters. Here we are considering the time-space separation. For the sake of simplicity, we assume that this loading consists of only a normal force distribution $f(t)$ over the upper surface of the substructure (see Figure 31).

Figure 32 shows the evolution of the constitutive relation error e_{CR}^2 as-

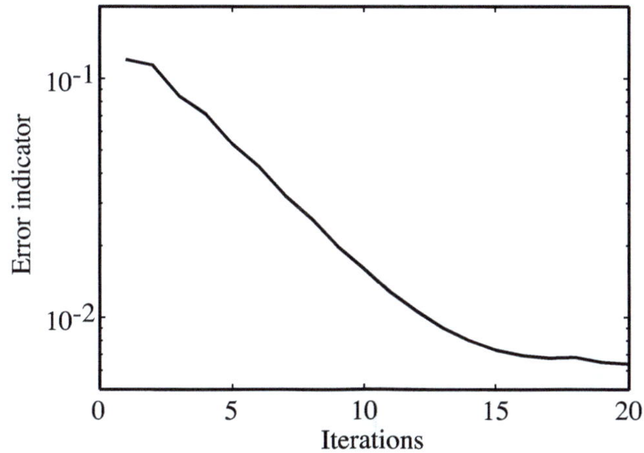

Figure 29. Convergence of the method

sociated with the search direction versus the number of functions, obtained using two techniques: in the first technique, new pairs of time/space functions were built systematically; the second technique consisted, as described in Algorithm 1, in first reusing the reduced basis of space functions calculated previously to update the time functions alone, and only after that seeking a new pair of time/space functions. One can see that the accuracy of the approximation is very good because with only 4 radial functions the error was less than 1%. However, one can observe that the convergence rate was higher with the second technique than with the first. For example, in order to get less than 0.1% error, 15 functions needed to be calculated without updating the time functions, as opposed to only 8 functions using the updating procedure. Since the computation cost associated with reusing the reduced basis is much less than the cost associated with an additional space function, it is very important to update the time functions systematically.

Figure 33, shows the first four pairs consisting of a space function and a time function. Since the space functions are normalized, one can observe a decrease in the level of the corresponding time functions. Figure 34 shows a comparison of the Von Mises stress distributions over the space and time domains obtained with the radial time-space approximation and the classical incremental technique.

A very important point is that although the basis of the space functions is *a priori* specific to the problem and the loading for which it was defined, it can be reused to solve another problem with comparable accuracy. For

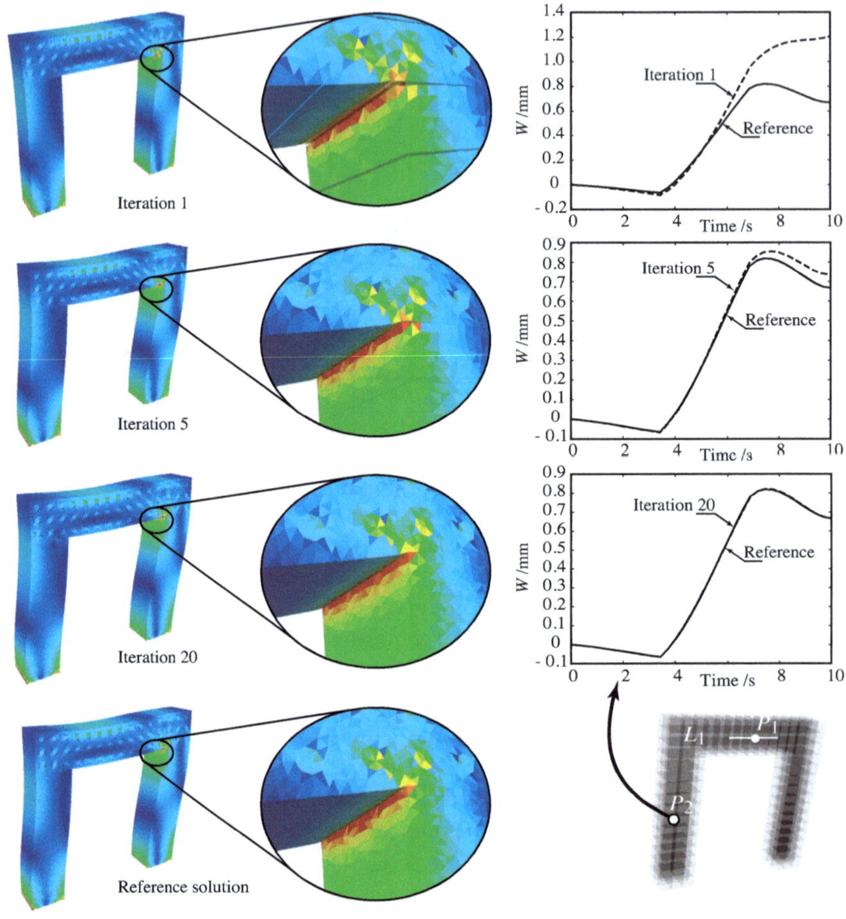

Figure 30. Approximate solutions after various numbers of iterations

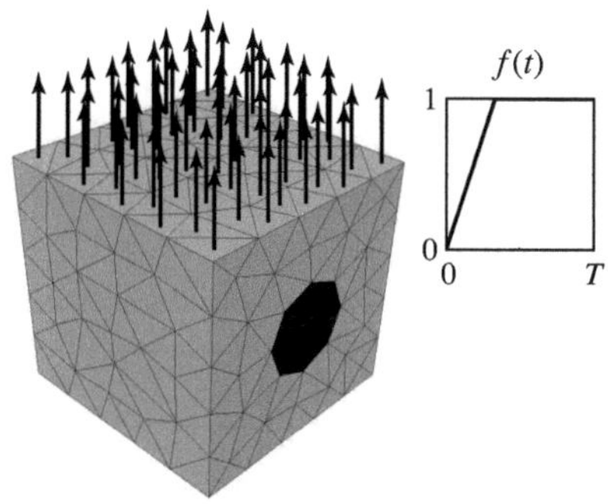

Figure 31. Description of a micro problem corresponding to a Type-II substructure and its loading

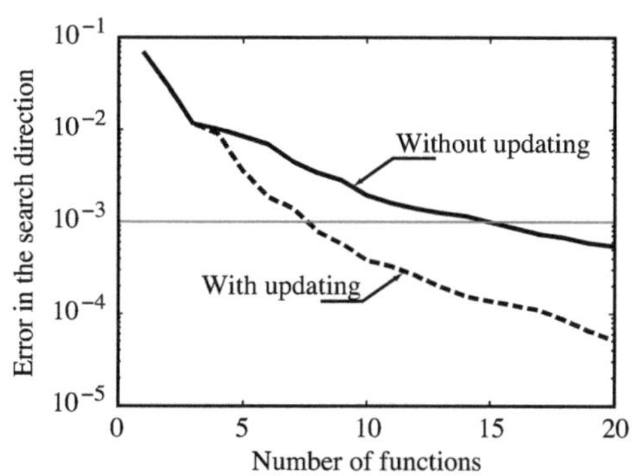

Figure 32. Convergence of the PGD approximation

example, we solved the previous micro problem with 6 functions and reused these functions for all the loading cases of Figure 35. In order to do that, we carried out a single updating stage and evaluated the corresponding error.

Table 36 shows that by updating the time functions alone using the same space functions as for a previous problem $f(t)$ one obtains an approximate solution of the new problem $f_i(t)$ with an accuracy comparable to that of the first problem. The robustness of the PGD approximation makes it well-adapted to multiresolution and, thus, this approximation technique is quite suitable for the multiscale strategy, which involves the resolution of a set of micro problems at each iteration of the LATIN method. We can reuse the same basis at every iteration of these micro problems, and even consider using a common basis for the whole set of substructures.

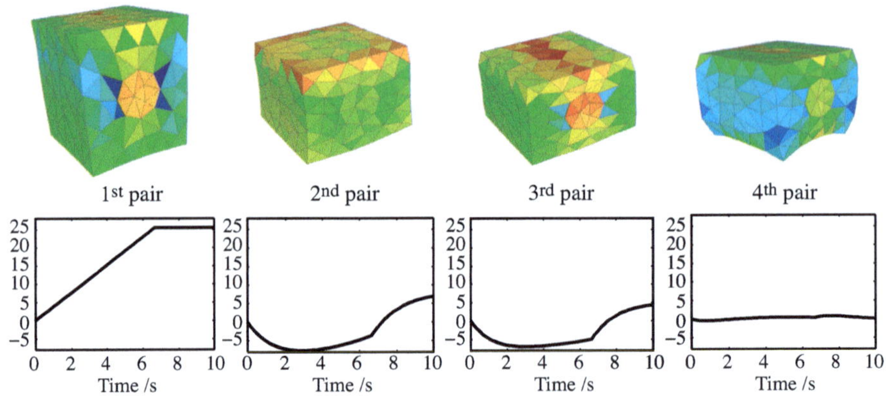

Figure 33. The first four radial time-space functions for the problem

Remark

The lagrange multiplier $\underline{\dot{\widetilde{W}}}_E^M$ could be seen as parameters in the PGD framework. That as been done in (Cremonesi et al., 2013) and leads to an important gain in term of computational cost.

3.6 The Reference Point Method (RPM)

Working with PGD description alone constitues a very convenient framework in which the storage requirement is drastically reduced. However, one has to consider product of PGD descriptions and integrals of these products whose cost is relatively high. Several methods have been proposed to overcome theses difficulties. One should mention the Empirical Interpolation Method (EIM) (Barrault et al., 2004) and its discrete version named DEIM (Chaturantabut and Sorensen, 2010). Here, we are developing the Reference Point Method introduced in (Ladevèze, 1997) and developed with D. Néron and M. Capaldo.

Definition of the reference points

Let us divide the time interval I being studied into m subintervals $\{I_i\}_{i=1,\ldots,m}$ of lengths $\{\Delta t_i\}_{i=1,\ldots,m}$ as shown in Figure 37. Introducing the centers $\{t_i\}_{i=1,\ldots,m}$ of these subintervals, called "reference times", one has $I_i = [t_i - \Delta t_i/2, t_i + \Delta t_i/2]$.

In the space domain, let us also introduce m' points $\{\underline{M}_j\}_{j=1,\ldots,m'}$ and

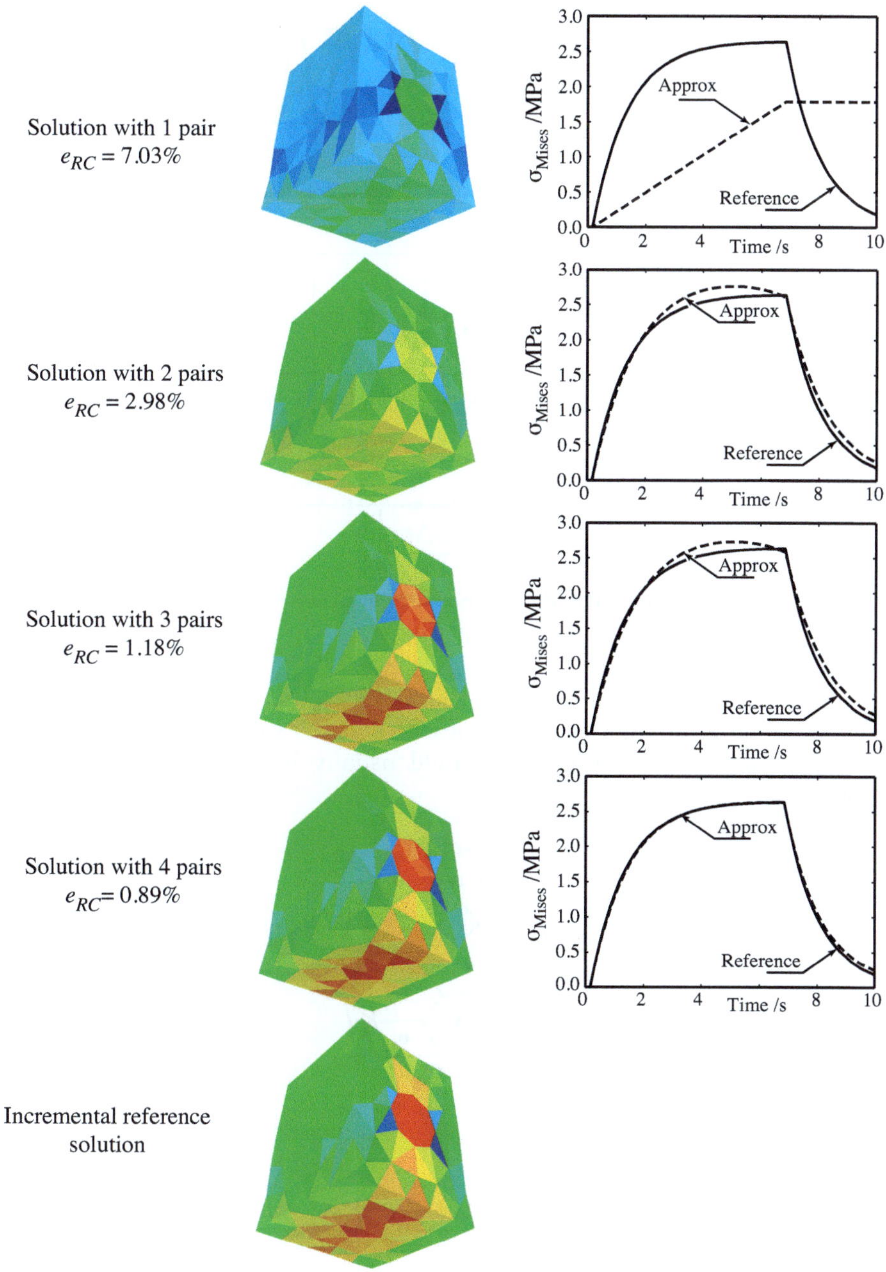

Figure 34. Quality of the approximations with 1, 2, 3 and 4 pairs

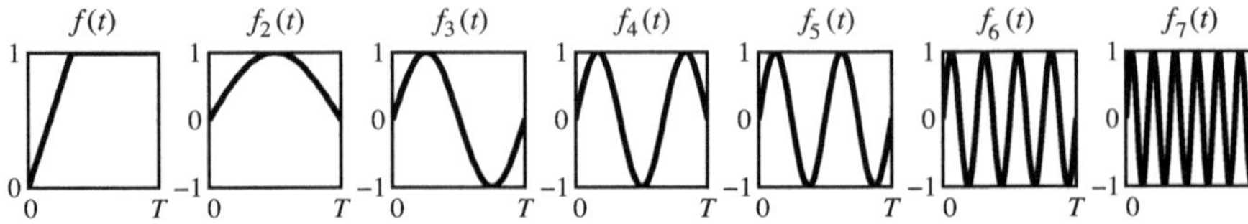

Figure 35. The different loading cases for the micro problem

Loading case	$f(t)$	$f_2(t)$	$f_3(t)$	$f_4(t)$	$f_5(t)$	$f_6(t)$	$f_7(t)$
Error e_{CR}	0.179%	0.183%	0.239%	0.291%	0.332%	0.411%	0.434%

Figure 36. Reutilization of a space function with the PGD approximation

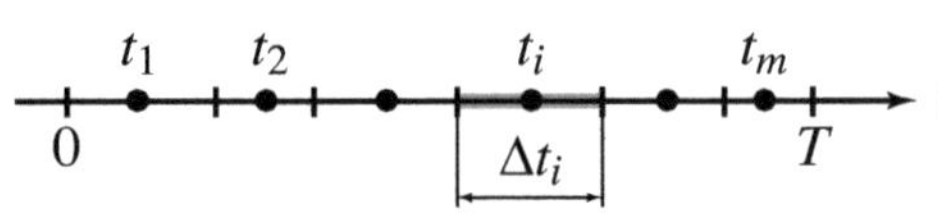

Figure 37. The reference times in I

partition Ω into $\{\Omega_j\}_{j=1,\dots,m'}$ as shown in Figure 38. These points are called "reference points" and the measures of the subdomains are denoted $\{\omega_j\}_{i=j,\dots,m'}$. In practice, there would usually be a few dozen reference points.

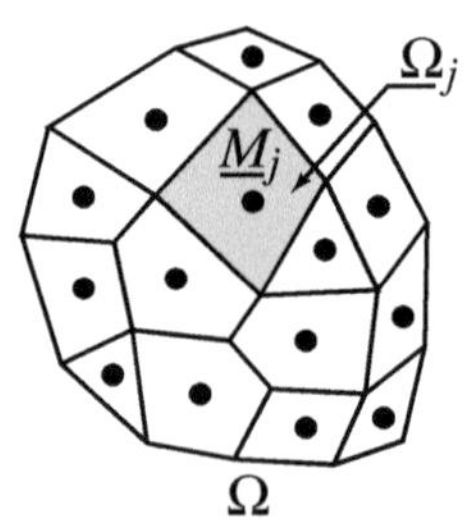

Figure 38. The reference points in Ω

The choice of these reference times and reference points is unrelated to the classical discretizations of the time interval I and space domain Ω. Refined time and space discretizations should still be used for the calculation of the various quantities. Here, our purpose is to describe a field f over the time-space domain $I \times \Omega$ through:

$$\hat{a}_i^j(t) = \begin{cases} f(t, \underline{M}_j) & \text{if } t \in I_i \\ 0 & \text{otherwise} \end{cases} \quad \text{and} \quad \hat{b}_i^j(\underline{M}) = \begin{cases} f(t_i, \underline{M}) & \text{if } \underline{M} \in \Omega_j \\ 0 & \text{otherwise} \end{cases} \tag{73}$$

with $i = 1, \ldots, m$ and $j = 1, \ldots, m'$. The sets $\{(\hat{a}_i^j, \hat{b}_i^j)\}_{i=1,\ldots,m}^{j=1,\ldots,m'}$ are the generalized components of f. One should note that these quantities verify the following compatibility conditions: for $i = 1, \ldots, m$ and $j = 1, \ldots, m'$,

$$\hat{a}_i^j(t_i) = \hat{b}_i^j(\underline{M}_j) \tag{74}$$

It could arrive that the quantity f is not well represented over the time-space domain. Then, one adds if necessary a PGD description of the complement.

Remark

The extension to paramter-dependent functions is easy. Let us consider one parameter p belonging to Γ. The generalized components are:

$$\begin{aligned} \hat{a}_j^k(t) &= f(t, \underline{M}_j, p_k) && \text{if } t \in I \\ \hat{b}_i^k(M) &= f(t_i, \underline{M}, p_k) && \text{if } \underline{M} \in \Omega \\ \hat{c}_i^j(p) &= f(t_i, \underline{M}_j, p) && \text{if } p \in \Gamma \end{aligned} \tag{75}$$

Reconstruction

Then, the main question is: how can one build or rebuild a field from its components? We choose to define function f from its components using only one product per time-space subdomain $I_i \times \Omega_j$:

$$f(t, \underline{M}) : a_i^j(t) b_i^j(\underline{M}) \quad \forall (t, \underline{M}) \in I_i \times \Omega_j \tag{76}$$

where the sets $\{(a_i^j, b_i^j)\}_{i=1,\ldots,m}^{j=1,\ldots,m'}$ should be defined from the sets $\{(\hat{a}_i^j, \hat{b}_i^j)\}_{i=1,\ldots,m}^{j=1,\ldots,m'}$. However, in this case, we let the time domain play a special role because there are many more spatial degrees of freedom than there are time degrees of freedom. Thus, function f is defined by:

$$f(t, \underline{M}) : a_i(t) b_i(\underline{M}) \quad \forall (t, \underline{M}) \in I_i \times \Omega \tag{77}$$

Let us introduce the following scalar products:

$$\langle f, g \rangle_{I_i} = \int_{I_i} fg \, dt \quad \text{and} \quad \langle f, g \rangle_{\Omega_j} = \int_{\Omega_j} fg \, d\Omega \tag{78}$$

In order to obtain the sets $\{(a_i, b_i)\}_{i=1,\dots,m}$, we minimize:

$$J(a_i, b_i) = \sum_{j=1}^{m'} \left[\omega_j \|\hat{a}_i^j(t) - a_i(t) b_i(\underline{M}_j)\|_{I_i}^2 + \Delta t_i \|\hat{b}_i^j(\underline{M}) - a_i(t_i) b_i(\underline{M})\|_{\Omega_j}^2 \right] \tag{79}$$

which leads to:

$$a_i(t) = \frac{\sum_{j=1}^{m'} \omega_j \hat{a}_i^j(t) b_i(\underline{M}_j)}{\sum_{j=1}^{m'} \omega_j b_i^2(\underline{M}_j)} \quad \text{and} \quad b_i(\underline{M}) = \frac{\sum_{j=1}^{m'} \hat{b}_i^j(\underline{M})}{m' a_i(t_i)} \tag{80}$$

Consequently, $\forall (t, \underline{M}) \in I_i \times \Omega_j$, we obtain:

$$f(t, \underline{M}) : a_i(t) b_i(\underline{M}) = \frac{\sum_{k=1}^{m'} \omega_k \hat{a}_i^k(t) \hat{b}_i^k(\underline{M}_k)}{\sum_{k=1}^{m'} \omega_k \hat{b}_i^k(\underline{M}_k) \hat{b}_i^k(\underline{M}_k)} \hat{b}_i^j(\underline{M}) \tag{81}$$

Then, using the compatibility conditions (74), we get:

$$f(t, \underline{M}) : a_i(t) b_i(\underline{M}) = \frac{\sum_{k=1}^{m'} \omega_k \hat{a}_i^k(t) \hat{a}_i^k(t_i)}{\sum_{k=1}^{m'} \omega_k \hat{a}_i^k(t_i) \hat{a}_i^k(t_i)} \hat{b}_i^j(\underline{M}) \tag{82}$$

These formula are easily extended to the case where one has also parameters. It is clear that operations (addition, product, derivation) are greatly facilitated using the generalized components.

Application

The main point is the computation of the search direction defining the residue to minimize for getting the PGD approximation. Here, one does need a very accurate computation and then the Reference Point Method is used without additional PGD complement Comparison with EIM is in progress.

3.7 Prospects

Numerical simulation has made a forceful entry into design and analysis offices and it is only the beginning. This revolution as entered to a new stage, called "simulation-driven robust design", which poses some major scientific challenges. The main key are model order reduction methods with offline and online works, and in particular the PGD. Engineering applications as the one of paragraph 3.4 will be more and more numerous.

PGD has been developed for most of wild mechanics issues but robust PGD-computational methods are still missing for several important issues:

- large displacement problems with instabilities;
- damage and final fracture (in particular for composites);
- transient dynamics;
- large number of parameters.

In addition to these scientific challenges, the computation technique itself should be improved and the RPM seems to be promising.

Bibliography

S. Akel and Q. S. Nguyen. Determination of the limit response in cyclic plasticity. In *Proceedings of 2nd International Conference on Computational Plasticity*, volume 639-650, Barcelone, Spain, 1989.

O. Allix. Damage analysis of delamination around a hole. In P. Ladevèze and O.C. Zienkiewicz, editors, *New advances in computational structural mechanics*. Elsevier, 1992.

O. Allix and P. Vidal. A new multi-solution approach suitable for structural identification problems. *Computer Methods in Applied Mechanics and Engineering*, 191:2727–2758, 2002.

A. Ammar, F. Chinesta, P. Diez, and A. Huerta. An error estimator for separated representations of highly multidimensional models. *Computer Methods in Applied Mechanics and Engineering*, 199:1872–1880, 2010.

X. Aubard, P.-A. Boucard, P. Ladevèze, and S. Michel. Modeling and simulation of damage in elastomer structures at high strains. *Computers & Structures*, 80:2289–2298, 2002.

A. Barbarulo, P. Ladevèze, H. Riou, and L. Kovalesky. Proper generalized decomposition applied to linear acoustic: a new tool for broad band calculation. *Journal of Sound and Vibration*, To appear, 2014.

M. Barrault, Y. Maday, N.C. Nguyen, and A.T. Patera. An "empirical interpolation" method: Application to efficient reduced-basis discretization of partial differential equations. *Comptes Rendus Académie des Sciences Paris*, 339:667–672, 2004.

T. Belytschko, P. Smolinski, and W. K. Liu. Stability of multi-time step partitioned integrators for first-order finite element systems. *Computer Methods in Applied Mechanics and Engineering*, 49(3):281–297, 1985.

P Boisse, P Bussy, and P Ladevèze. A new approach in nonlinear mechanics - the large time increment method. *International Journal For Numerical Methods In Engineering*, 29(3):647–663, 1990.

C. L. Bottasso. Multiscale temporal integration. *Computer Methods in Applied Mechanics and Engineering*, 191(25-26):2815–2830, 2002.

P.-A. Boucard and P. Ladevèze. A multiple solution method for non-linear structural mechanics. *Mechanical Engineering*, 50(5):317–328, 1999.

P.-A. Boucard, P. Ladevèze, M. Poss, and P. Rougée. A nonincremental approach for large displacement problems. *Computers & Structures*, 64: 499–508, 1997.

S. Boyaval, C. Le Bris, Y. Maday, N.C. Nguyen, and A.T. Patera. A reduced basis approach for variational problems with stochastic parameters: application to heat conduction with variable robin coefficient. *Computer Methods in Applied Mechanics and Engineering*, 198(41-44):3187–3206, 2009.

P. Bussy, P. Rougée, and P. Vauchez. The large time increment method for numerical simulation of metal forming processes. In *NUMETA*, pages 102–109. Elsevier, 1990.

L. Chamoin and P. Ladevèze. Bounds on history-dependent or independent local quantities in viscoelasticity problems solved by approximate methods. *International Journal for Numerical Methods in Engineering*, 71(12):1387–1411, 2007.

L. Chamoin and P. Ladevèze. A non-intrusive method for the calculation of strict and efficient bounds of calculated outputs of interest in linear viscoelasticity problems. *Revue Européenne de Mécanique Numérique*, 17:981–992, 2008.

L. Chamoin and P. Ladevèze. Robust control of pgd-based numerical simulations. *European Journal of Computational Mechanics*, 21(3-6):195–207, 2012.

L. Chamoin, P. Ladevèze, and F. Pled. Goal oriented control of finite element models: recent advances and performances on 3D industrial applications. *Design and Analysis*, 1:107–118, 2012.

Saifon Chaturantabut and Danny C. Sorensen. Nonlinear model reduction via discrete emprical interpolation. *Siam Journal On Scientific Computing*, 32(5):2737–2764, 2010. doi: DOI 10.1137/090766498.

F. Chinesta. Proper generalized decomposition in extreme simulations: Towards a change of paradigm in computational mechanics? *IACM Expressions*, 2009.

J.-Y. Cognard and P. Ladevèze. A large time increment approach for cyclic viscoplasticity. *International Journal of Plasticity*, 9:141–157, 1993.

JY Cognard, P Ladeveze, and P Talbot. A large time increment approach for thermo-mechanical problems. *Advances In Engineering Software*, 30 (9-11):583–593, Sep-Nov 1999.

A. Combescure and A. Gravouil. A numerical scheme to couple subdomains with different time-steps for predominantly linear transient analysis. *Computer Methods in Applied Mechanics and Engineering*, 191: 1129–1157, April 2002.

F. Comte, M. H. Maitournam, P. Burry, and T. Mac Lan Nguyen. A direct method for the solution of evolution problems. *Comptes Rendus Mecanique*, 334(5):317–322, 2006.

M. Cremonesi, D. Néron, P.-A. Guidault, and P. Ladevèze. A PGD-based homogenization technique for the resolution of nonlinear multiscale problems. *Computer Methods in Applied Mechanics and Engineering*, 267: 275–292, 2013. doi: http://dx.doi.org/10.1016/j.cma.2013.08.009.

F. Devries, F. Dumontet, G. Duvaut, and F. Léné. Homogenization and damage for composite structures. *International Journal for Numerical Methods in Engineering*, 27:285–298, 1989.

D. Dureisseix and C. Farhat. A numerically scalable domain decomposition method for the solution of frictionless contact problems. *International Journal for Numerical Methods in Engineering*, 50:2643–2666, 2001.

D. Dureisseix, P. Ladevèze, D. Néron, and B. A. Schrefler. A multi-time-scale strategy for multiphysics problems: application to poroelasticity. *International Journal for Multiscale Computational Engineering*, 1(4): 387–400, 2003a.

D Dureisseix, P Ladevèze, and BA Schrefler. A latin computational strategy for multiphysics problems: application to poroelasticity. *International Journal For Numerical Methods In Engineering*, 56(10):1489–1510, 2003b. doi: DOI 10.1002/nme.622.

C. Farhat and M. Chandesris. Time-decomposed parallel time-integrators: theory and feasibility studies for fluid, structure, and fluid-structure applications. *International Journal for Numerical Methods in Engineering*, 58:1397–1434, 2003.

V. Faucher and A. Combescure. A time and space mortar method for coupling linear modal subdomains and non-linear subdomains in explict structural dynamics. *Computer Methods in Applied Mechanics and Engineering*, 192:509–533, 2003.

F. Feyel. A multilevel finite element (FE^2) to describe the response of highly non-linear structures using generalized continua. *Computer Methods in Applied Mechanics and Engineering*, 192:3233–3244, 2003.

J. Fish and W. Chen. Uniformly valid multiple spatial-temporal scale modeling for wave propagation in heterogeneous media. *Mechanics of Composite Materials and Structures*, 8:81–99, 2001.

J. Fish, K. Shek, M. Pandheeradi, and M. S. Shephard. Computational plasticity for composite structures based on mathematical homogenization: Theory and practice. *Computer Methods in Applied Mechanics and Engineering*, 148:53–73, 1997.

G. H. Golub and C. F. Van Loan. *Matrix computations*. Johns Hopkins University Press, Baltimore, 3nd edn edition, 1996.

A. Gravouil and A. Combescure. Multi-time-step explicit implicit method for non-linear structural dynamics. *International Journal for Numerical Methods in Engineering*, 50:199–225, 2001.

A. Gravouil and A. Combescure. Multi-time-step and two-scale domain decomposition method for non-linear structural dynamics. *International Journal for Numerical Methods in Engineering*, 58:1545–1569, 2003.

M.A. Grepl and A.T. Patera. A posteriori error bounds for reduced-basis approximation of parametrized parabolic partial differential equations. *ESAIM-Mathematical Modelling and Numerical Analysis (M2AN)*, 39(1):157–181, 2005.

T. Guennouni. On a computational method for cycling loading: the time homogenization. *Mathematical Modelling and Numerical Analysis (in french)*, 22(3):417–455, 1988.

Ch. Heyberger, P.-A. Boucard, and D. Néron. A rational strategy for the resolution of parametrized problems in the PGD framework. *Computer Methods in Applied Mechanics and Engineering*, 259:40–49, 2013.

Christophe Heyberger, Pierre-Alain Boucard, and David Néron. Multiparametric analysis within the proper generalized decomposition framework. *Computational Mechanics*, 49(3):277–289, 2012.

C. Huet. Application of variational concepts to size effects in elastic heterogeneous bodies. *Journal of the Mechanics and Physics of Solids*, 38(6): 813–841, 1990.

T. J. R. Hughes. Multiscale phenomena: Green's function, the Dirichlet-to-Neumann formulation, subgrid scale models, bubbles and the origin of stabilized methods. *Computer Methods in Applied Mechanics and Engineering*, 127:387–401, 1995.

V.G. Kouznetsova, M.G.D. Geers, and W.A.M. Brekelmans. Multi-scale constitutive modelling of heterogeneous materials with a gradient-enhanced computational homogenization scheme. *International Journal for Numerical Methods in Engineering*, 54:1235–1260, 2002.

P. Ladevèze. Strict upper error bounds for computed outputs of interest in computational structural mechanics. *Computational Mechanics*, 42(2): 271–286, 2008.

P. Ladevèze. Upper error bounds on calculated outputs of interest for linear and nonlinear structural problems. *Comptes Rendus Académie des Sciences Paris*, 334(7):399–407, 2006.

P. Ladevèze. New algorithms: mechanical framework and development (in french). Technical Report 57, LMT-Cachan, 1985a.

P. Ladevèze. On a family of algorithms for structural mechanics (in french). *Compte rendu de l'académie des Sciences*, 300(2):41–44, 1985b.

P. Ladevèze. The large time increment method for the analyse of structures with nonlinear constitutive relation described by internal variables. *Comptes Rendus Académie des Sciences Paris*, 309(II):1095–1099, 1989.

P. Ladevèze. New advances in the large time increment method. In P. Ladevèze and O. C. Zienkiewicz, editors, *New advances in computational structural mechanics*, pages 3–21. Elsevier, 1991.

P. Ladevèze. *Mécanique non linéaire des structures*. Hermès, Paris, 1996.

P. Ladevèze. A computational technique for the integrals over the time-space domain in connection with the LATIN method (in french). Technical Report 193, LMT-Cachan, 1997.

P. Ladevèze. *Nonlinear Computationnal Structural Mechanics—New Approaches and Non-Incremental Methods of Calculation*. Springer Verlag, 1999.

P. Ladevèze and L. Chamoin. Calculation of strict error bounds for finite element approximations of nonlinear pointwise quantities of interest. *International Journal for Numerical Methods in Engineering*, 84:1638–1664, 2010.

P. Ladevèze and L. Chamoin. On the verification of model reduction methods based on the proper generalized decomposition. *Computer Methods in Applied Mechanics and Engineering*, 200:2032–2047, 2011.

P. Ladevèze and L. Chamoin. *Toward guaranteed PGD-reduced models*. Bytes and Science, G. Zavarise & D.P. Boso (Eds.), CIMNE, 2012.

P Ladevèze and D Dureisseix. A new micro-macro computational strategy for structural analysis. *Comptes Rendus De L'Academie Des Sciences Serie Ii Fascicule B-Mecanique Physique Astronomie*, 327(12):1237–1244, 1999.

P. Ladevèze and D. Leguillon. Error estimate procedure in the finite element method and application. *SIAM Journal of Numerical Analysis*, 20(3): 485–509, 1983.

P. Ladevèze and Ph. Lorong. A large time increment approach with domain decomposition technique for mechanical non linear problems. *Computing Methods in Applied Sciences and Engineering INRIA*, pages 569–578, 1992.

P. Ladevèze and E.A.W Maunder. A general method for recovering equilibrating element tractions. *Computer Methods in Applied Mechanics and Engineering*, 137:111–151, 1996.

P. Ladevèze and Ph. Maurel. Numerical implementation of a new method for the analysis of 3D structures. Technical report, Aerospatiale TKS/S 47541 (in french), 1990.

P. Ladevèze and A. Nouy. On a multiscale computational strategy with time and space homogenization for structural mechanics. *Computer Methods in Applied Mechanics and Engineering*, 192:3061–3087, 2003.

P. Ladevèze and J.P. Pelle. *Mastering Calculations in Linear and Nonlinear Mechanics.* Springer NY, 2004.

P Ladevèze and U Perego. Duality preserving discretization of the large time increment methods. *Computer Methods In Applied Mechanics and Engineering*, 189(1):205–232, 2000.

P. Ladevèze and P. Rougée. (Visco)plasticity under cyclic loadings: properties of the homogenized cycle. *Comptes Rendus de l'Académie des Sciences*, II(301):891–894, 1985.

P. Ladevèze, P. Marin, J.P. Pelle, and J.L. Gastine. Accuracy and optimal meshes in finite element computation for nearly incompressible materials. *Computational Methods in Applied Mechanical Engineering*, 94(3):303–314, 1992.

P. Ladevèze, J.-Y. Cognard, and P. Talbot. A non-incremental and adaptive computational approach in thermo-viscoplasticity. In O. T. Bruhns and E. Stein, editors, *IUTAM Symposium on Micro- and Macrostructural Aspects of the Thermoplasticity*, pages 281–291, 1999.

P. Ladevèze, O. Loiseau, and D. Dureisseix. A micro-macro and parallel computational strategy for highly heterogeneous structures. *International Journal for Numerical Methods in Engineering*, 52:121–138, 2001.

P. Ladevèze, A. Nouy, and O. Loiseau. A multiscale computational approach for contact problems. *Computer Methods in Applied Mechanics and Engineering*, 191:4869–4891, 2002.

P. Ladevèze, D. Néron, and P. Gosselet. On a mixed and multiscale domain decomposition method. *Computer Methods in Applied Mechanics and Engineering*, 196:1526–1540, 2007.

P. Ladevèze, D. Néron, and J.-C. Passieux. On multiscale computational mechanics with time-space homogenization. In *Multiscale methods—Bridging the scales in Science and Engineering*, chapter Space Time Scale Bridging methods, pages 247–282. Oxford University Press. Ed. J. Fish, 2009.

P. Ladevèze, L. Chamoin, and E. Florentin. A new non-intrusive technique for the construction of admissible stress fields in model verification. *Computer Methods in Applied Mechanics and Engineering*, 199(9-12): 766–777, 2010.

P. Ladevèze, J.-C. Passieux, and D. Néron. The LATIN multiscale computational method and the Proper Generalized Decomposition. *Computer Methods in Applied Mechanics and Engineering*, 199:1287–1296, 2010.

P. Ladevèze, B. Blaysat, and E. Florentin. Strict upper bounds of the error in calculated outputs of interest for plasticity problems. *Computational Methods in Applied Mechanical Engineering*, 245:194–205, 2012.

P. Ladevèze, F. Pled, and L. Chamoin. New bounding techniques for goal-oriented error estimation applied to linear problems. *International Journal for Numerical Methods in Engineering*, 93(13):1345–1380, 2013.

M. Lefik and B.A. Schrefler. Modelling of nonstationary heat conduction problems in micro-periodic composites using homogenisation theory with corrective terms. *Archives of Mechanics*, 52(2):203–223, 2000.

M. H. Maitournam, B. Pommier, and J.-J. Thomas. Determination de la reponse asymptotique d'une structure anelastique sous chargement thermomecanique cyclique. *Comptes Rendus Mecanique*, 330(10):703–708, 2002.

D. Néron. A wavelet-based multiscale approach for the resolution of time-dependent PDEs in the proper generalized decomposition framework. In *Proceedings of the Fourth International Conference on Computational Methods for Coupled Problems in Science and Engineering*, 2011.

D. Néron and D. Dureisseix. A computational strategy for poroelastic problems with a time interface between coupled physics. *International Journal for Numerical Methods in Engineering*, 73(6):783–804, 2008a.

D. Néron and D. Dureisseix. A computational strategy for thermo-poroelastic structures with a time-space interface coupling. *International Journal for Numerical Methods in Engineering*, 75(9):1053–1084, 2008b. doi: 10.1002/nme.2283.

D. Néron and P. Ladevèze. Proper Generalized Decomposition for multiscale and multiphysics problems. *Archives of Computational Methods in Engineering*, 17(4):351–372, 2010.

D. Néron and P. Ladevèze. A data compression approach for PGD reduced-order modeling. In *Proceedings of the 11th Biennial Conference On Engineering Systems Design And Analysis, ASME*, 2012.

A. Nouy. A priori model reduction through proper generalized decomposition for solving time dependent partial differential equations. *Computer Methods in Applied Mechanics and Engineering*, 199:1603–1626, 2010.

A. Nouy. A generalized spectral decomposition technique to solve a class of linear stochastic partial differential equations. *Computer Methods in Applied Mechanics and Engineering*, 196(45-48):4521–4537, 2007.

A. Nouy and P. Ladevèze. Multiscale computational strategy with time and space homogenization: a radial type approximation technique for solving micro problems. *International Journal for Multiscale Computational Engineering*, 170(2):557–574, 2004.

J. T. Oden, K. Vemaganti, and N. Moës. Hierarchical modeling of heterogeneous solids. *Computer Methods in Applied Mechanics and Engineering*, 172:3–25, 1999.

J-P. Pelle and D. Ryckelynck. An efficient adaptive strategy to master the global quality of viscoplastic analysis. *Computers & Structures*, 78(1-3): 169–184, 2000.

F. Pled, L. Chamoin, and P. Ladevèze. On the techniques for constructing admissible stress fields in model verification: performances on engineering examples. *International Journal for Numerical Methods in Engineering*, 88(5):409–441, 2011.

F. Pled, L. Chamoin, and P. Ladevèze. An enhanced method with local energy minimization for the robust a posteriori construction of equilabrated stress fields in finite elements analysis. *Computational Mechanics*, 47(3): 357–378, 2012.

N. Relun, D. Néron, and P.-A. Boucard. A model reduction technique based on the PGD for elastic-viscoplastic computational analysis. *Computational Mechanics*, 51(1):83–92, 2013.

D.V. Rovas, L. Machiels, and Y. Maday. Reduced-basis output bound methods for parabolic problems. *IMA Journal of Numerical Analysis*, 26(3): 423–445, 2006.

E. Sanchez-Palencia. Comportement local et macroscopique d'un type de milieux physiques hétérogènes. *International Journal of Engineering Science*, 12(4):331–351, 1974.

E. Sanchez-Palencia. Non homogeneous media and vibration theory. *Lecture Notes in Physics*, 127, 1980.

J. Waeyetens, L. Chamoin, and P. Ladevèze. Guaranteed error bounds on pointwise quantities of interest for transient viscodynamics problems. *Computational Mechanics*, 49(3):291–307, 2012.

O.C. Zienkiewicz and R.L. Taylor. *The finite element method, volume 1.* MPG Books Ltd, 2000.

T. Zohdi and P. Wriggers. *Introduction to computational micromechanics.* Springer Verlag, 2005.

T.I. Zohdi, J.T. Oden, and G.J. Rodin. Hierarchical modeling of heterogeneous bodies. *Computer Methods in Applied Mechanics and Engineering*, 138(1-4):273–298, 1996.

Fundamentals of reduced basis method for problems governed by parametrized PDEs and applications

Gianluigi Rozza*
SISSA, International School for Advanced Studies, *mathLab*
Via Bonomea 265, 34136 Trieste, Italy
gianluigi.rozza@sissa.it

Abstract

In this chapter we consider Reduced Basis (*RB*) approximations of parametrized Partial Differential Equations (*PDEs*). The the idea behind *RB* is to decouple the generation and projection stages (Offline/Online computational procedures) of the approximation process in order to solve parametrized *PDEs* in a fast, inexpensive and reliable way. The *RB* method, especially applied to 3D problems, allows great computational savings with respect to the classical Galerkin Finite Element (FE) Method. The standard *FE* method is typically ill suited to (i) iterative contexts like optimization, sensitivity analysis and many-queries in general, and (ii) real time evaluation. We consider for simplicity coercive *PDE*s. We discuss all the steps to set up a *RB* approximation, either from an analytical and a numerical point of view. Then we present an application of the *RB* method to a steady thermal conductivity problem in heat transfer with emphasis on geometrical and physical parameters.

1 Introduction

Any design problem can be summarized in the following statement: *Given a physical process and given a set of suitable* parameters, *the physical be-*

*This work has been developed thanks to the contribution of the SISSA NOFYSAS excellence grant and INdAM-GNCS. Mr Alberto Trezzini is kindly acknowledged for the preparation of the schemes and for the numerical results. G. Rozza ackwnowledges the long lasting collaboration with Prof. A.T. Patera (MIT) and his group on the subject. This Chapter is based on the lectures given by the author in summer 2013 at CISM, International Centre for Mechanical Sciences in Udine, Italy. Thanks to Prof. G. Maier (Politecnico di Milano) for insights and remarks.

F. Chinesta, P. Ladevèze (Eds.), *Separated Representations and PGD-Based Model Reduction Fundamentals and Applications,* CISM International Centre for Mechanical Sciences
DOI 10.1007/ 978-3-7091-1794-1_4 © CISM Udine 2014

haviour depends on finding the optimal values of these parameters in order to obtain a desirable behaviour of the process.
In this chapter the attention is focused on the designer point of view, but it is also important to recall that there are many other applications of interest, such as *real time* evaluation of the performances of a system or *sensitivity analysis* with respect to certain parameters.
The *physical process* may belong to any field of engineering interest, such as, heat and mass transfer, elasticity, acoustics, fluid dynamics, electromagnetism, etc., or even in a broader sense to any quantitative disciplines (e.g., finance, biology, ecology and medicine) and their interdisciplinary combinations.
The physical process is analysed in order to find the best mathematical *model* able to describe the behaviour of the *system*. As *model* we refer to a system of equations and/or other mathematical relationships able to "catch" the main properties of the process and to predict its evolution in time and/or space. In this stage the engineer introduces all the simplifications that the observation and a subsequent qualitative analysis suggest to take into account.
The analytical model is then constituted by conservation laws and constitutive equations. In this work the attention will be focused on models whose conservation laws are those of *continuum mechanics* which appears as balance of suitable quantities (e.g., mass, energy, linear momentum, angular momentum, etc...).
The constitutive relationships come from experimental evidence and depends upon the essential features of the process itself. The result of the combination of these conservation laws and constitutive equations is often an equation or a system of *Partial Differential Equations* PDEs. This means that in the equations the unknowns will appear along with partial derivatives with respect to multiple variables (temporal or spatial). To solve differential equation it is also necessary, in order to obtain the closure of the problem, specify a suitable set of boundary/initial conditions.
Therefore the parameters in the process can be:

- physical, within this category we have:
  - coefficient of constitutive equation for the particular physical process addressed;
  - non-dimensional numbers;[1]
  - imposed boundary conditions;
- and geometrical.

[1]It is always necessary to write the *PDEs* in a non-dimensional version in order to highlight the actual physical dependence on these parameters.

The desired behaviour could be a particular performance of the system, such as an average temperature in a thermal block, a maximum displacement of a loaded beam, a level of vorticity in a flow-field and so on, depending on the particular process addressed.
Finally, the optimal configuration of these parameters can be found through an iterative optimization process in which a suitable *cost functional*, that depends upon the particular desired performance, has to be minimized.
To wit, the designer is interested to evaluate iteratively the *input-output* relationship to estimate the cost functional.
The solution of *PDEs* by classical discretization methods like finite element, spectral method or finite volume, typically involves thousands (in some cases millions) of degrees of freedom (DOFs) to obtain a "good" solution; therefore a single evaluation of the input-output relationship is very expensive and at last, in most cases, not suitable in a *many-query* context on which the design strategy is based on.
In this context it is necessary to rely on suitable *Reduced Order Modelling* ROM techniques that reduce the computational cost and time. The *Reduced Basis* (RB) method is one of the possible options and this memoir focuses on this method that, as it will be shown, is able to reduce the computational cost of *orders of magnitude*. Moreover the *RB* method is a *certified* and *reliable* method because in addition to an efficient output calculation is able to provide *rigorous a-posteriori error estimators* on the *"truth" solution*[2].
An important remark is constituted by the fact that all the procedure is built upon a *reference domain* on which a *suitable discretization* is constructed. Then there is no need to re-discretize the domain at each iteration, re-building a mesh, or deforming the domain itself in the case of a *parameter-dependent geometry.*

1.1 Overview on reduced basis method

As already said the *RB* method is a *reduced order method* that is able to reduce the complexity of a system without loss of information or accuracy of the results thanks to the *rigorous error bound* provided and by the properties of the *Galerkin* projection (Patera and Rozza, 2007; Rozza et al., 2008).
This method does not replace an existing discretization method but "works in collaboration with it" and upon it.
In this work the discretization method adopted for the applications is the *finite element method* (FE): this choice does not constitute a limitation be-

[2]The discussion on our assumption of "truth" solution will be given in section1.2.

cause the *RB* method is built upon the *user-defined* assumption of *"truth"* solution. The choice of the correct method able to describe the physical process is demanded to the user and it will not be considered here.
The basic idea is to start with *FE* basis of dimension $\mathcal{N}$ and then construct a *RB* basis whose dimension N is much smaller than the former, so that $N << \mathcal{N}$.
The power of the *RB* method resides in the splitting of the computational procedures into two parts:

1. an Offline phase,
2. an Online phase,

where the former is $\mathcal{N}$ dependent and computed once, whereas the latter is N dependent and allows a *fast*, *inexpensive* and *reliable* input-output evaluation.
The role played by this decomposition is immediately clear taking in consideration for example an optimization process. The optimal configuration can be found thanks to an iterative process, in which a *PDEs* state solution is needed at every step of iteration. In this context the *RB* advantage is that the evaluation of the solution for every step is orders of magnitude smaller; furthermore the real time evaluation of the *PDEs* solution is feasible, unlike as in the *FE* case.
The splitting procedure is possible if the weak formulation of the *PDEs* can be expressed in an *affine parameter decomposition.* This is one of the *key-point* of the procedure: all the parametric dependences of the *PDEs* are actually separated by the non-parametric part, in order to allow computing the latter just in a *reference domain*, whereas the former can be computed several times with a very inexpensive computational burden (Rozza, 2009b). A graphical sketch of this idea is shown in Figure 1. The *RB* method is depicted in the upper figure , whereas the *FE* classical method is depicted in the lower one. In the *RB* sketch it appears that the *Offline* computational burden that is proportional to K$\mathcal{N}$, where $\mathcal{N}$ is the number of degrees of freedom for the *FE* problem and K is a positive constant that depends upon many ingredients that will be in depth discussed in section 3. It is worth to recall that in the *RB* methodology the so-called *Offline* part is very expensive but in a *real time* or in a *many-query context* the most important part is played by the *online* evaluation of the *input-output relationship* which is very inexpensive in the *RB* methodology.
In addition, by dealing with parametrized equations it is clear that in many cases it is possible to restrict the attention on a tipically *smooth* and rather *low-dimensional parametrically induced manifold* of the functional space, which is a set of fields engendered as the input varies over the parameter domain (Patera and Rozza, 2007; Rozza et al., 2008). In Section 1.2 a more

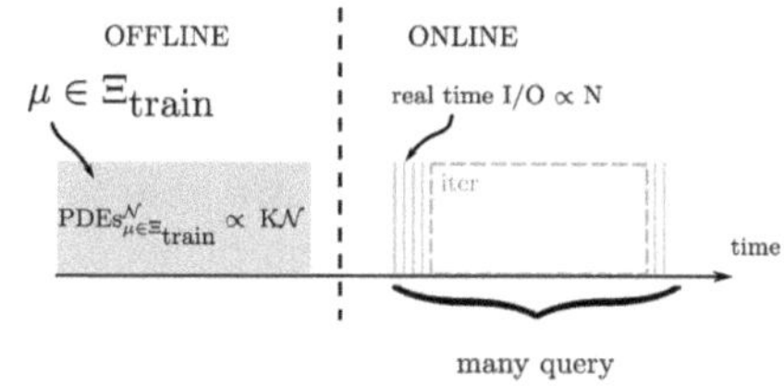

(a) *RB* method

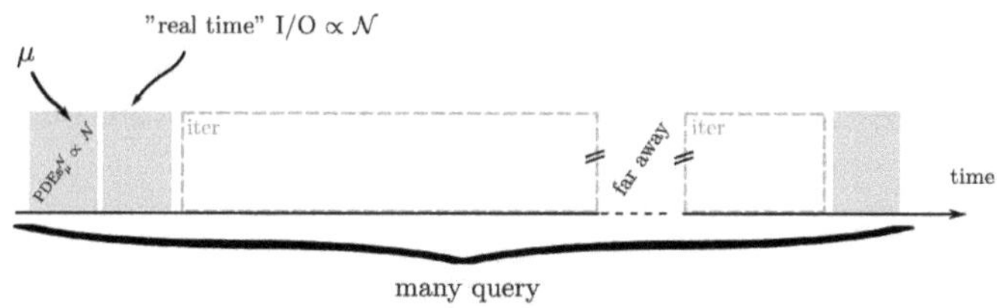

(b) *FE* method

Figure 1: *RB* computational saving

involved discussion over these concepts will be given, which are the key points the reduced basis method has been built upon.

After the motivations and scopes of the *RB* method, a brief historical perspective will be recalled in this section, then in section 2 we will recall some mathematical generalities and we will provide the abstract formulation for coercive problems.
In section 3 the steps for the generation of the *RB* approximation spaces for the solution of parametrized *PDEs* will be explained. Later, in section 4, the *affine geometry preconditions* will be presented focused on the 3D case, necessary to allow a fully decoupling between *Offline* and *Online* procedures.
In section 5 the *RB* approximation for a parametrized 3D thermal block will be addressed, followed by some conclusions and perspectives.

1.2 RB scope and historical perspective

In the past few years, thanks to the increased computational performances it has been possible to use numerical simulation in the very first steps of design in several fields.
Unfortunately despite this hardware improvement, the greater part of engi-

neering problem involves the solution of partial differential equations, furthermore in a *design context* the number of solutions for various configurations of attempt can become *very large* and eventually excessive. Therefore it is necessary to develop techniques that are able to reduce the complexity of the system without a loss of information or accuracy of the results. The *RB* method is a promising approach to respond to this need, moreover this method is not only *rapid* and *efficient*, but also provides a reliable solution of partial differential equations thanks to a certified *a-posteriori error bound*.
This method provides a useful tool for engineers, in fact thanks to the very low cost of the input-output relationship evaluation, the design procedure can be enriched with highly accurate numerical simulation since the very first steps.

The *real-time* and *many-query contexts* represent not only computational challenges but also computational opportunities.
It is possible to identify two key opportunities that can be gainfully exploited with *RB* method (Rozza et al., 2008):

- **Opportunity I**
  In the parametric setting, the attention is restricted to a typically *smooth* and rather *low-dimensional parametrically induced manifold*: the set of fields engendered as the input varies over the parameter domain; in the case of single parameter, the parametrically induced manifold is a one-dimensional filament within the infinite dimensional space which characterizes general solutions to the PDEs. Clearly, generic approximation spaces are *unnecessarily rich* and hence *unnecessarily expensive* within the parametric framework.
- **Opportunity II**
  In the *real-time* or *many-query contexts*, in which the premium is on marginal cost (or equivalently asymptotic average cost) per *input-output* evaluation, we can accept greatly increased pre-processing or *Offline* cost, not admitted for a single or few evaluations, in exchange for greatly decreased *Online* (or deployed) cost for each new/additional input-output evaluation (Rozza et al., 2008). Clearly, resources allocation typical for *single-query* investigations will be far from optimal for many-query and real-time exercises. We shall review the development of RB methods in terms of these two opportunities.

Opportunity I: A reduced basis discretization is, in brief, a Galerkin projection on an N-dimensional approximation space that focuses on the parametrically induced manifold identified in Opportunity I.

Initial work (about in 1980s) grew out of two related streams of inquiry: from the need for more effective, and perhaps also more interactive, many-query design evaluation (Fox and Miura, 1971) considering linear structural examples; and from the need of more efficient parameter continuation methods (Almroth et al., 1978; Noor and Peters, 1980; Noor, 1978, 1982) considering nonlinear structural analysis problems.

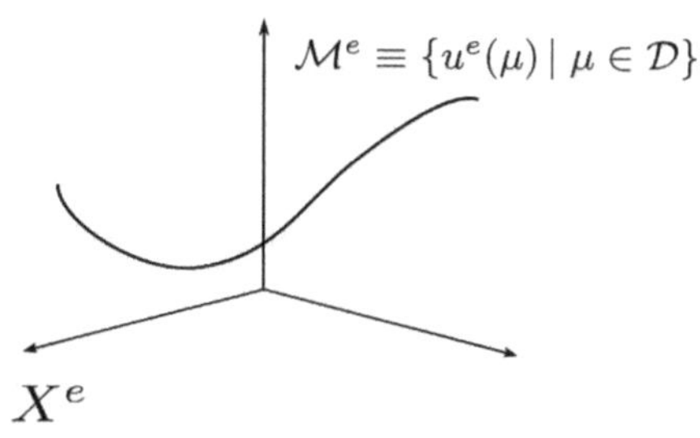

(a) X^e space

Figure 2: Parametrically induced manifold

Some modal analysis techniques proposed in the same years (Nagy, 1979) deals with geometrically nonlinear behaviour and are closely related to RB notions.
At the very first moment the reduced basis method arises from the study of nonlinear elasticity problems. The development of the *RB* method has been mainly due to the engineering needs to obtain a very efficient tool in the design context.
The following decade saw further expansion into different applications and classes of equations (Ito and Ravindran, 1998) dealing also with the control of PDEs and (Peterson, 1989) applications of reduced order methods to fluid dynamics and incompressible Navier-Stokes equations. However, in these early methods, the approximation spaces tended to be rather local and typically rather low dimensional in parameter (often say a single parameter). In part, this was due to the nature of the applications taken in account (parametric continuation), but it was also due to the *absence* of *a-posteriori error estimators* and *effective sampling procedures*. In fact the absence of this kind of techniques did not allow a *certified* an *accurate* prediction of the error between the *"truth" solution*[3] and the solution obtained by the reduced order model; the lack of a rigorous error certification is unacceptable for example in a safe engineering context, in which the reliability is an imperative.
Much current effort is thus devoted to development of (i) a posteriori error estimation procedures, in particular rigorous error bounds for outputs of interest (Prud'homme et al., 2002a) and (ii) effective sampling strategies, in particular for higher dimensional parameter domains (Bui-Thanh et al.,

[3] "Truth" here means the assumption of faithfulness achieved with the standard discretization method, see section 2.3 for further explanation on the hypothesis and on the consequences of this choice

2008; Nguyen et al., 2005; Rozza, 2009a).
The *a-posteriori error bounds* are of course mandatory for *rigorous certification* of any particular reduced basis (*Online*) output evaluation. However, the error estimators can also play an important role in efficient and effective (greedy) sampling procedures: the inexpensive error bounds permit us, first, to explore much larger subsets of the parameter domain in search of most representative or best *snapshots*, and, second, to determine when the basis functions are enough to control the error within a certified bound.
The most used sampling methods are (i) the *Greedy sampling procedure* and (ii) the *POD* (Proper Orthogonal Decomposition): these two procedures differ under some aspects that will be discussed in section 3.3. It is worth to anticipate that the *Greedy* procedure is optimized for higher dimensions of the parameter space, while the *POD* procedure is better suited for one dimensional (typically time) domain (Atwell and King, 1999; Kunisch and Volkwein, 2003; Ravindran, 2002; Willcox and Peraire, 2002).

Opportunity II: Early work on the reduced basis method took into consideration the Opportunity II, but without being able to fully decouple the underlying standard (*FE*) discretization.
More precisely, often the *Galerkin stiffness equations* for the reduced basis system were generated by direct appeal to the high dimensional *FE* representation: essentially, pre and post multiplying the FE stiffness system by rectangular basis matrices; as a result the computational saving offered by the model reduction was not fully exploited Noor (1978); Porsching (1985); Porsching and Lee (1987). The complete decoupling between the reduced order model and the standard discretization model is one of the crucial point on which much of the current work has been devoted at that time.
In this work we will denote with $\mathcal{N}$ the computational complexity of the standard discretization and with N the *RB* complexity. This opportunity has been exploited thanks to an *Offline/Online procedure* that is possible in a context of an *affine parameter dependence* of the operators constituting the *PDEs*; this important concept will be recalled in section 2.2.
One of the benefits related to this procedure is shown in Figure 3; it is clear that in a many-query context, such as an optimization design process, or in a real-time closed loop control the split allows a much more faster subiteration (*I/O* evaluation), that could not be possible in the classical *FE* discretization. In fact in the FE case (Figure 3b) a subiteration involves a cost proportional to $\mathcal{N}$, whereas RB involves a cost proportional only to N.
The Offline-Online idea is quite self-apparent and it has been treated quite often (Ito and Ravindran, 1998; Jabbar and Azeman, 2004; Peterson, 1989), nonetheless the idea/application of *a-posteriori error bound* is more involved

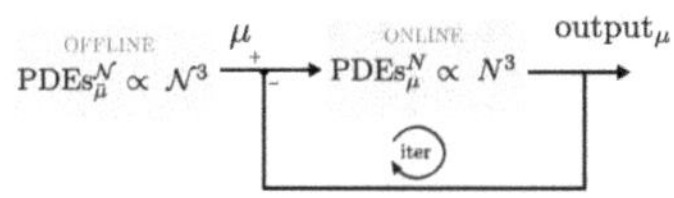

(a) RB procedure

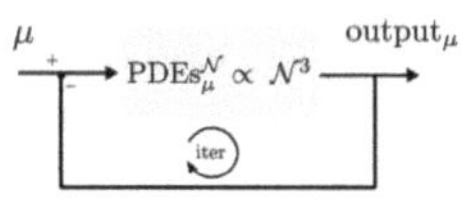

(b) FE procedure

Figure 3: Offline/Online splitting

and recent (Huynh et al., 2007; Prud'homme et al., 2002a,b).
Actually even in the case of a *non-affine parameter dependence* the development of the procedure is yet possible, although more complex procedures, that have been estabilished in the few last years (M. et al., 2004; Grepl et al., 2007), are needed in order to turn the non-affine form into an approximated affine problem.

1.3 Problems addressed

This chapter will deal with *linear output functional* and *affinely parametrized linear coercive* PDEs: these problems are relevant to many important applications in various field of engineering interest.
Although we focuse on these problems, the reduced basis method is much more general and is able to treat nonaffine problems (Grepl et al., 2007) and parabolic equations (Atwell and King, 1999).
Furthermore, the *RB* method may be used for nonlinear equations such as the incompressible (quadratically non-linear) *Navier-Stokes* equations, finally even the hyperbolic equations are subject of study, there are proofs which demonstrate that RB approximation and a posteriori error estimation can be applied, although up to now there are still many issues related to *smoothness* and *stability* (Haasdonk and Ohlberger, 2008; Patera and Ronquist, 2007).
The application chosen in this work deals with thermal conduction in continuum mechanics and it proves to be a convenient expository vehicle for the methodology exploitation.

2 Parametrized Elliptic PDEs

In the first part of this section, we will briefly introduce some generalities about *parametric* bilinear form, *parametric* linear functional, coercivity and continuity constants. In particular the abstract formulation for coercive problem will be reported.

2.1 Parametric operators

In this section, definitions and properties about *parametric bilinear* and *bilinear forms* will be introduced.
The theory presented here is available with further details and explanations in Patera and Rozza (2007) for the coercive case, in Rozza and Veroy (2006) for the non coercive case.
The basic concept of functional analysis concerning Cartesian product, functional norms, bilinear forms and dual spaces are given as known (Quarteroni and Valli, 1997; Yosida, 1971).

Linear and bilinear parametric forms. We first introduce a closed bounded parameter domain $\mathcal{D} \subset \mathbb{R}^P$ with a typical parameter vector, or $P-$tuple, in $\mathcal{D}$ shall be denoted $\boldsymbol{\mu} = (\mu_1, \ldots, \mu_P)$. We assume that $\mathcal{D}$ is suitably regular.
It is now necessary to introduce some definition that resembles the classical definition for non-parametric linear operators.

Definition 2.1. Let Z be an inner product space over $\mathbb{R}$, $b : Z \times Z \times \mathcal{D} \to \mathbb{R}$ is a *parametric bilinear form* if, for all $\boldsymbol{\mu} \in \mathcal{D}$, $b(\,\cdot\,,\,\cdot\,;\boldsymbol{\mu}) : Z \times Z \to \mathbb{R}$ is a bilinear form, i.e. for any $\alpha \in \mathbb{R}$ and for any $w, v, z \in Z$:

$$b(\,\alpha w + v,\, z\,;\boldsymbol{\mu}) = \alpha b(\,w,\, z\,;\boldsymbol{\mu}) + b(\,v,\, z\,;\boldsymbol{\mu}) \quad \forall\, \boldsymbol{\mu} \in \mathcal{D}$$

and for any $\beta \in \mathbb{R}$ and for any $w, v, z \in Z$:

$$b(\,w,\, \beta v + z\,;\boldsymbol{\mu}) = \beta b(\,w,\, v\,;\boldsymbol{\mu}) + b(\,w,\, z\,;\boldsymbol{\mu}) \quad \forall\, \boldsymbol{\mu} \in \mathcal{D}.$$

Definition 2.2. A parametric bilinear form $b : Z \times Z \times \mathcal{D} \to \mathbb{R}$is *symmetric* if, for any $v, w \in Z$:

$$b(\,w,\, v\,;\boldsymbol{\mu}) = b(\,v,\, w\,;\boldsymbol{\mu}) \quad \forall\, \boldsymbol{\mu} \in \mathcal{D} \tag{1}$$

A parametric bilinear form $b : Z \times Z \times \mathcal{D} \to \mathbb{R}$ is *skew-symmetric* if, for any $v, w \in Z$:

$$b(\,w,\, v\,;\boldsymbol{\mu}) = -b(\,v,\, w\,;\boldsymbol{\mu}) \quad \forall\, \boldsymbol{\mu} \in \mathcal{D} \tag{2}$$

Starting from the Definition 2.2 it is possible to define the symmetric and the skew-symmetric part of the bilinear form as it follows:

Definition 2.3. Given a parametric bilinear form, we define:

- the symmetric part of b as:

$$b_S(w, v; \boldsymbol{\mu}) = \frac{1}{2}\left(b(w, v; \boldsymbol{\mu}) + b(v, w; \boldsymbol{\mu})\right) \tag{3}$$

- the skew-symmetric part of b as:

$$b_{SS}(w, v; \boldsymbol{\mu}) = \frac{1}{2}\left(b(w, v; \boldsymbol{\mu}) - b(v, w; \boldsymbol{\mu})\right). \tag{4}$$

For the coercive case it is necessary to introduce the definition of the coercivity of the parametric bilinear form.

Coercivity.

Definition 2.4. We say that a parametric bilinear form $b : Z \times Z \times \mathcal{D} \to \mathbb{R}$ is *coercive* over Z if:

$$\alpha(\boldsymbol{\mu}) = \inf_{w \in Z} \frac{b(w, w; \boldsymbol{\mu})}{||w||_Z^2} \tag{5}$$

is *positive* for all $\boldsymbol{\mu} \in \mathcal{D}$.

We can define $(0 <)\ \alpha_0 = \min_{\boldsymbol{\mu} \in \mathcal{D}} \alpha(\boldsymbol{\mu})$.

Continuity. Now it is possible to define the *continuity* of the parametric bilinear form in very similar way. We say that:

Definition 2.5. A parametric bilinear form $b : Z \times Z \times \mathcal{D} \to \mathbb{R}$ is *continuous* over Z if:

$$\gamma(\boldsymbol{\mu}) = \sup_{w \in Z} \sup_{v \in Z} \frac{b(w, v; \boldsymbol{\mu})}{||w||_Z ||v||_Z} \tag{6}$$

is *finite* for all $\boldsymbol{\mu} \in \mathcal{D}$.

It is useful to define $\gamma_0 = \max_{\boldsymbol{\mu} \in \mathcal{D}} \gamma(\boldsymbol{\mu})$.

Linear parametric form. Similarly as already done with the parametric bilinear form, we recall:

Definition 2.6. Let Z be an inner product space over $\mathbb{R}$. $g : Z \times \mathcal{D} \to \mathbb{R}$ is a *parametric linear form* if, for all $\boldsymbol{\mu} \in \mathcal{D}$, and for any $w \in Z$, $g(\,\cdot\,;\boldsymbol{\mu}) : Z \times \mathcal{D} \to \mathbb{R}$ is a linear form. That is, for all $\alpha \in \mathbb{R}$, and for any $w, v \in Z$:

$$g(\,\alpha w + v\,;\boldsymbol{\mu}) = \alpha g(\,w\,;\boldsymbol{\mu}) + g(\,v\,;\boldsymbol{\mu}) \quad \forall\, \boldsymbol{\mu} \in \mathcal{D} \tag{7}$$

Continuity. In order to ensure the well posedness of the problem, the continuity of the parametric linear form is needed. We say that:

Definition 2.7. A parametric linear form g is *continuous* if, for all $\boldsymbol{\mu} \in \mathcal{D}$, $g(\,\cdot\,;\boldsymbol{\mu}) \in Z'$.

Z' denotes the *dual space*, that we recall is the space of all linear bounded functionals over Z. Note that the dual norm of a parametric linear form g, $||g(\,\cdot\,;\boldsymbol{\mu})||_{Z'}$, will of course be a (finite) function of $\boldsymbol{\mu}$ over $\mathcal{D}$.

Coercivity eigenproblem. We recall here an additional problem which will be useful in order to evaluate a rigouros error bound as it will be seen in the section 3.5 (Patera and Rozza, 2007; Quarteroni et al., 2000).
It is necessary to introduce an eigenproblem because in the subsequent analysis it will be useful to recognize that $\alpha(\boldsymbol{\mu})$ (and $\beta(\boldsymbol{\mu})$ in the non-coercive case) can be seen as the minimum eigenvalue of a *generalized eigenproblem.*

It is possible to rewrite (5) replacing the form b with his symmetric part, denoted with b_s:

$$\alpha(\boldsymbol{\mu}) = \inf_{w \in Z} \frac{b_s(\,w,\,w\,;\boldsymbol{\mu})}{||w||_Z^2} \tag{8}$$

it follows that $\alpha(\boldsymbol{\mu})$ can be expressed as a minimun eigenvalue.
It is useful to introduce the coercivity symmetric (generalized) eigenproblem associated with the parametric bilinear form $b : Z \times Z \times \mathcal{D} \to \mathbb{R}$.
Given $\boldsymbol{\mu} \in \mathcal{D}$, find the couple $(\chi,\,\lambda)_i(\boldsymbol{\mu}) \in Z \times \mathbb{R}$, $i = 1, \ldots, \dim(Z)$, such that:

$$b_S(\chi_i,\,v;\,\boldsymbol{\mu}) = \nu_i(\chi_i(\boldsymbol{\mu}), v)_Z \tag{9}$$

and

$$||\chi_i(\boldsymbol{\mu})|| = 1 \tag{10}$$

of course it will be possible to sort the $\dim(Z)$ eigenvalues in ascending order such that: $\lambda_1(\boldsymbol{\mu}) < \ldots < \lambda_{\dim(Z)}(\boldsymbol{\mu})$.

It simply descends from (9) and (8) that if b is coercive, then $\alpha(\boldsymbol{\mu}) = \lambda_1(\boldsymbol{\mu}) > 0$.

2.2 Affine parametric dependence

The *affine parametric dependence* of the bilinear form and of the linear functional is one of the most important ingredient in the offline/online decomposition and of course in the real-time input/output evaluation.
The idea is rather simple: split all the parametric dependent component by those parametrically independent. In addition we recall also the parametric coercivity definition.

Affine parametric bilinear forms. With regard to the bilinear parametric forms the affine dependence states that:

Definition 2.8. A parametric bilinear form $b : Z \times Z \times \mathcal{D} \to \mathbb{R}$ is *affine* in the parameter $\boldsymbol{\mu}$ if, for all $v, w \in Z$:

$$b(w, v; \boldsymbol{\mu}) = \sum_{q=1}^{Q_b} \theta_b^q(\boldsymbol{\mu})\, b^q(w, v) \quad \forall\, \boldsymbol{\mu} \in \mathcal{D} \tag{11}$$

for some finite, preferably small, Q_b.

Here the $\theta_b^q(\boldsymbol{\mu}) : \mathcal{D} \to \mathbb{R}$ are (tipically very smooth) parameter- dependent functions, and $b^q(w, v) : Z \times Z \to \mathbb{R}$ are parameter-independent bilinear forms.

Parametric coercivity. In the scope of an affine dependence it is useful also to consider the parametric coercivity of the bilinear form.

Definition 2.9. We say that an affine parametric (coercive) form $b : Z \times Z \times \mathcal{D} \to \mathbb{R}$ (definition 2.8) is *parametrically coercive* if, $c \equiv b_s$ (the symmetric part of b) admits an *affine development*:

$$c(w, v; \boldsymbol{\mu}) = \sum_{q=1}^{Q_c} \theta_c^q(\boldsymbol{\mu})\, c^q(w, v) \quad \forall\, \boldsymbol{\mu} \in \mathcal{D} \tag{12}$$

that satisfies two conditions:

$$\theta_c^q(\boldsymbol{\mu}) > 0 \quad \forall w, v \in Z,\ \forall\, \boldsymbol{\mu} \in \mathcal{D},\ 1 \le q \le Q_b \tag{13}$$

and

$$c^q(v, v) > 0 \quad \forall v \in Z,\ 1 \le q \le Q_b. \tag{14}$$

Note that we suppose that each $c^q(w,v)$ is symmetric.

Affine parametric linear form. Similarly as done for the bilinear form it is worth to introduce the affine dependence for a linear bounded functional.

Definition 2.10. We shall say that the parametric linear form $g : Z \times \mathcal{D} \to \mathbb{R}$ is affine in the parameter if, for any $v \in Z$:

$$g(v; \boldsymbol{\mu}) = \sum_{q=1}^{Q_g} \theta_g^q(\boldsymbol{\mu})\, g^q(v) \quad \forall\, \boldsymbol{\mu} \in \mathcal{D} \tag{15}$$

for some finite Q_g.

Once again, here the $\theta_g^q : Z \times \mathcal{D} \to \mathbb{R}$ are smooth parameter-dependent functions and the $g^q(v) : Z \to \mathbb{R}$ are parameter-independent bounded linear forms.
The affine representations (11) and (15) are not unique, though in general will exist minimum Q_b (in the former case) and Q_g (in the latter) terms of expansion able to describe the forms with an affine development.
Tipically the number of terms Q mainly depends on the complexity of the parameter-dependent geometry. This concept will be treated in the section 4.3 where the decomposition in the 2D and in the more involved 3D case will be considered.

2.3 Abstract formulation of a coercive parametrized problem

In this section, an abstract problem for *coercive* elliptic partial differential equations with affine parameter dependence will be introduced. First the exact formulation (in weak form) of the problem will be presented, then a finite element discretization will be introduced in order to build the *"truth"* space on which the reduced basis will be built upon.
As mentioned in the introduction, the interest resides in the evaluation of the *solution field* and the *output* that depends on the state equation which is solution of a *PDE.*

Exact formulation. Let $\Omega \in \mathbb{R}^d$, $d = 1, 2, 3$ be a suitable physical domain with Lipschitz continuous boundary $\partial\Omega$. Let $\mathcal{D} \subset \mathbb{R}^P$ be the parameter domain.
Moreover let Γ, be a boundary measurable segments of $\partial\Omega$, over which we shall ultimately impose *Dirichlet boundary condition* on the components of

the field variable. We next introduce a suitable scalar space Y_i^e, $1 \le i \le d$:

$$Y_i^e \equiv \left\{ v \in H^1(\Omega) \mid v\, |_{\Gamma_i^D} = 0 \right\} \tag{16}$$

in general $H_0^1(\Omega) \subset Y^e \subset H^1(\Omega)$. Clearly if $\Gamma = \partial\Omega$, then $X^e \equiv H_0^1$.
We then construct the space in which our *vector-valued* field variable shall reside as a Cartesian product:

$$X^e = Y_1^e \times Y_2^e \times \ldots \times Y_1^d$$

We equip X^e with an inner product $(v, w)_{X^e}$, $\forall\, v, w \in X^e$ and induced norm $||w||_{X^e} = \sqrt{(w,w)_{X^e}}$, $\forall\, w \in X^e$; any inner product which induces a norm equivalent to the $\left(H^1\right)^d$ is admissible.

Problem statement. It is now possible to state the problem in the "exact" space: let $a : Z \times Z \times \mathcal{D} \to \mathbb{R}$ be a *continuous coercive parametric bilinear* form, let $f : Z \times \mathcal{D} \to \mathbb{R}$ and l be a *parametric linear functional bounded over* X^e.
Given $\boldsymbol{\mu} \in \mathcal{D} \subset \mathbb{R}^P$, find $u(\boldsymbol{\mu}) \in X^e$ such that

$$a\left(u^e(\boldsymbol{\mu}, v; \boldsymbol{\mu})\right) = f\left(v; \boldsymbol{\mu}\right) \quad \forall\, v \in X^e \tag{17}$$

and evaluate

$$s^e(\boldsymbol{\mu}) = l(u^e(\boldsymbol{\mu}); \boldsymbol{\mu}). \tag{18}$$

Here $s^e(\boldsymbol{\mu})$ is the output of interest, $s^e(\boldsymbol{\mu}) : \mathcal{D} \to \mathbb{R}$ is the input-output relationship and l is the linear *"output"* functional that links the input to the output through the field variable.
It follows from our hypothesis on a, f and l that the problem has a unique solution thanks to the *Lax-Milgram theorem* (Quarteroni et al., 2000; Quarteroni and Valli, 1997).
Recalling the affine development of the bilinear form and of the linear functional (section 2.2), it is possible to write the operators in the following form; for any $\boldsymbol{\mu} \in \mathcal{D}$:

$$a(w, v; \boldsymbol{\mu}) = \sum_{q=1}^{Q_a} \theta_a^q(\boldsymbol{\mu})\, a^q(w, v) \quad \forall\, v, w \in X^e \tag{19}$$

$$f(v; \boldsymbol{\mu}) = \sum_{q=1}^{Q_f} \theta_f^q(\boldsymbol{\mu})\, f^q(v) \quad \forall\, v \in X^e \tag{20}$$

$$l(v; \boldsymbol{\mu}) = \sum_{q=1}^{Q_l} \theta_l^q(\boldsymbol{\mu})\, l^q(v) \quad \forall\, v \in X^e \tag{21}$$

for finite and *preferebly* small Q_a, Q_f, Q_l. We implicitly assume that the θ_a^q for $1 \leq q \leq Q_a$, θ_f^q for $1 \leq q \leq Q_f$ and θ_l^q for $1 \leq q \leq Q_l$ are simple algebraic expressions that can be readily evaluated in $\mathcal{O}(1)$ operations.

Compliant problem. In this section the problems considered will be *"compliant"*, i.e. :

1. $l(\cdot\,;\boldsymbol{\mu}) = f(\cdot\,;\boldsymbol{\mu})$
2. $a(w;v;\boldsymbol{\mu}) = a(v,w;\boldsymbol{\mu}) \quad \forall\, w, v \in X^e$

that is the *output* functional and the *load/source* functional are the same and the bilinear form is symmetric (e.g. "compliancy" in linear elasticity). Considering these two hypotheses the problem (17)-(18) can be rewritten as it follows.
Given $\boldsymbol{\mu} \in \mathcal{D} \subset \mathbb{R}^P$, find $u(\boldsymbol{\mu}) \in X^e$ such that

$$a\left(u^e(\boldsymbol{\mu}, v; \boldsymbol{\mu})\right) = f\left(v; \boldsymbol{\mu}\right) \quad \forall\, v \in X^e \tag{22}$$

and evaluate

$$s^e(\boldsymbol{\mu}) = f(u^e(\boldsymbol{\mu}); \boldsymbol{\mu}). \tag{23}$$

Truth approximation. We focus the attention on the *"truth" approximation.* The reduced basis approximation will be built upon and the error will be measured relatives to this assumption of truth.
The role played by this assumption is very important; during the basis assembling and the error measuring the *RB* will completely "forget" the error between the *exact* solution and the *truth-assumption.* Then it is necessary to take some caution in order to ensure that this error remains suitably small for any given $\boldsymbol{\mu} \in \mathcal{D}$.
For analytical purposes, we assume that the *"truth"* takes the place of the *exact* statement. *Standard finite element FE* approximation (Quarteroni, 2013) may be chosen to represent the truth and to measure the error in order to build the *RB* basis and evaluate the error bounds for a given new set of parameter input $\boldsymbol{\mu}$.

Galerkin projection. We introduce a family of conforming approximation spaces $X^{\mathcal{N}} \subset X^e$ of dimension $\dim(X^{\mathcal{N}}) = \mathcal{N} < \infty$.
We then associate to our space a set of basis functions $\phi_k^{\mathcal{N}} \subset X^{\mathcal{N}}$, $1 \leq k \leq \mathcal{N}$, by construction, any member of $X^{\mathcal{N}}$ can be represented by a unique linear combination of the basis functions $\phi_k^{\mathcal{N}} \subset X^{\mathcal{N}}$.
Finally, we associate the inner products and induced norms $X^{\mathcal{N}}$ is equipped, denoted by $(v,w)_{X^{\mathcal{N}}}$, $\forall\, v, w \in X^{\mathcal{N}}$, and induced norm $||w||_{X^{\mathcal{N}}} = \sqrt{(w,w)_{X^{\mathcal{N}}}}$, $\forall\, w \in X^{\mathcal{N}}$.

This inner product, along with those related to the exact space, is explained below.

Inner product and induced norms. We now define the inner product and the norm over the space $X^{\mathcal{N}}$ and X^e and the *energy norm* given by the coercive bilinear form a.
For $w, v \in X^e$, we define respectively the *energy inner product* and the *energy norm* as:

$$\begin{aligned} ((w,v))_{\boldsymbol{\mu}} &= a\left(w,v;\boldsymbol{\mu}\right), \\ |||w|||_{\boldsymbol{\mu}} &= \sqrt{(w,w)_{\boldsymbol{\mu}}}, \end{aligned} \tag{24}$$

moreover, for a given $\bar{\boldsymbol{\mu}} \in \mathcal{D}$, we define for $w, v \in X^e$ the X^e-inner product and the X^e-norm as:

$$\begin{aligned} (w,v)_{X^e} &= ((w,v))_{\bar{\boldsymbol{\mu}}} + \tau\,(w,v)_{L^2(\Omega)}, \\ ||w||_{X^e} &= \sqrt{(w,w)_{X^e}}, \end{aligned} \tag{25}$$

where τ is a negative real parameter and$(w,v)_{L^2(\Omega)} = \displaystyle\int_\Omega w\,v\,d\Omega$.
Remark 1: We note that in order to define our X^e-norm we have chosen a fixed valued of the parameter $\overline{\boldsymbol{\mu}}$.
Remark 2: since $X^{\mathcal{N}} \subset X^e$, the inner products and the norms defined above are the same for the space $X^{\mathcal{N}}$.
The choice of $\bar{\boldsymbol{\mu}}$ and τ will affect the quality and efficiency of our reduced basis a posteriori error estimators, but this will not affect directly our reduced basis output predictions (Rozza et al., 2008).

Truth problem statement. Now we can state the problem in the truth space taking the *Galerkin projection* of the problem (22)-(23); given $\boldsymbol{\mu} \in \mathcal{D} \subset \mathbb{R}^P$, find $u(\boldsymbol{\mu}) \in X^{\mathcal{N}}$ such that

$$a\left(u^{\mathcal{N}}(\boldsymbol{\mu}), v; \boldsymbol{\mu}\right) = f\left(v;\boldsymbol{\mu}\right) \quad \forall\, v \in X^{\mathcal{N}} \tag{26}$$

and evaluate

$$s^{\mathcal{N}}(\boldsymbol{\mu}) = f(u^{\mathcal{N}}(\boldsymbol{\mu});\boldsymbol{\mu}). \tag{27}$$

Coercivity and continuity. We can define precisely the *exact* and the finite element *approximated coercivity constants* respectively, as:

$$\alpha^e(\boldsymbol{\mu}) = \inf_{w \in X^e} \frac{b(w, w; \boldsymbol{\mu})}{||w||^2_{X^e}}, \tag{28}$$

$$\alpha^{\mathcal{N}}(\boldsymbol{\mu}) = \inf_{w \in X^{\mathcal{N}}} \frac{b(w, w; \boldsymbol{\mu})}{||w||^2_{X^e}}. \tag{29}$$

From the coercivity hypothesis, we have that $\alpha^e(\boldsymbol{\mu}) > \alpha_0$, $\forall \boldsymbol{\mu} \in \mathcal{D}$; furthermore from our hypothesis on $X^{\mathcal{N}}$, that is a conforming space, we have that $\alpha^{\mathcal{N}}(\boldsymbol{\mu}) > \alpha^e(\boldsymbol{\mu})$, $\forall \boldsymbol{\mu} \in \mathcal{D}$. Than even after the approximation the problem remains coercive.

In the same way, the continuity constants are defined as

$$\gamma^e(\boldsymbol{\mu}) = \sup_{w \in X^e} \sup_{v \in X^e} \frac{b(w, v; \boldsymbol{\mu})}{||w||_{X^e} ||v||_{X^e}}, \tag{30}$$

$$\gamma^{\mathcal{N}}(\boldsymbol{\mu}) = \sup_{w \in X^{\mathcal{N}}} \sup_{v \in X^{\mathcal{N}}} \frac{b(w, v; \boldsymbol{\mu})}{||w||_{X^{\mathcal{N}}} ||v||_{X^{\mathcal{N}}}}, \tag{31}$$

once again from our hypothesis it follows $\gamma^e(\boldsymbol{\mu}) < \infty$ and $\gamma^{\mathcal{N}}(\boldsymbol{\mu}) \leq \gamma^e(\boldsymbol{\mu})$, $\forall \boldsymbol{\mu} \in \mathcal{D}$.

Well-posedness. The Galerkin approximation on $X^{\mathcal{N}}$ must satisfy the same conditions that the exact formulation satisfies over X^e. For the particular class of problems of interest in this section (elliptic coercive PDEs) the Galerkin formulation in fact *directly inherits* and even *improves upon* all the good properties of the exact formulation:

1. The dual norm of f over $X^{\mathcal{N}} (\subset X^e)$ is bounded by the dual norm of f over X^e;
2. symmetry is preserved;
3. a is coercive over $X^{\mathcal{N}}$ with:

$$\alpha^{\mathcal{N}}(\boldsymbol{\mu}) \geq \alpha^e(\boldsymbol{\mu}) \quad \forall \boldsymbol{\mu} \in \mathcal{D} \tag{32}$$

4. a is continuos over $X^{\mathcal{N}}$ with:

$$\gamma^{\mathcal{N}}(\boldsymbol{\mu}) \leq \gamma^e(\boldsymbol{\mu}) \quad \forall \boldsymbol{\mu} \in \mathcal{D} \tag{33}$$

5. the *affine expansions* for f and a are still valid for w, v restricted to $X^{\mathcal{N}}$;
6. a still satisfies the two conditions for parametric coercivity (section 2.2);

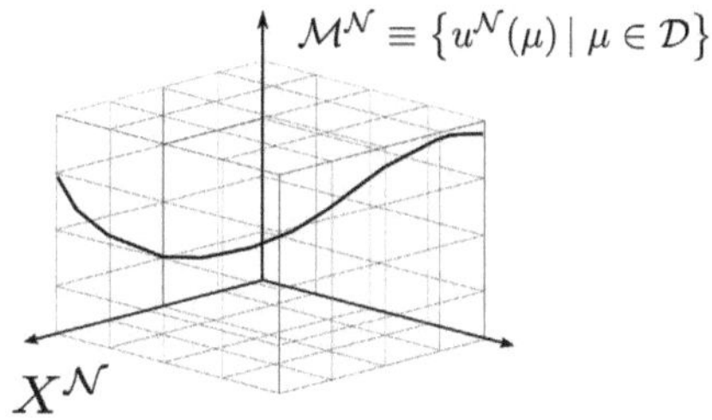

(a)

Figure 4: Parametrically induced manifold on $X^{\mathcal{N}}$

thus, for any $\mathcal{N}$ and associated $X^{\mathcal{N}}$, the Galerkin approximation preserves the *parametrically coercivity* and *affine compliancy* property.

3 RB method for parametrized elliptic coercive PDEs

In this section we will introduce the relevant steps for the generation of the rapidly convergent global RB approximation spaces for the approximation of the solution of parametrized *coercive* elliptic partial differential equations with affine parameter dependence will be explained.
Subsequently it will be possible to introduce the reduced basis approximation methodology, the sampling strategies and the construction of the reduced spaces.
Then an *a-posteriori error bound* necessary to achieve an efficient *RB* sampling it will be introduced.

3.1 The RB heuristic idea

As described in section 1.2, the *RB* approach derives from the two *opportunities*.
In particular regarding the *Opportunity I*, although $u^{\mathcal{N}}(\boldsymbol{\mu})$ is a member of the space $X^{\mathcal{N}}$ of typically very high dimension $\mathcal{N}$, in fact $u^{\mathcal{N}}(\boldsymbol{\mu})$ resides on a *low-dimensional parametrically induced manifold* $\mathcal{M} \equiv \{u^{\mathcal{N}}(\boldsymbol{\mu}) \mid \boldsymbol{\mu} \in \mathcal{D}\}$.
In Figure 4 a graphical heuristic idea of the finite dimensional (*truth*) manifold $X^{\mathcal{N}}$ with the parametrically induced manifold $\mathcal{M}^{\mathcal{N}}$ (filament) is shown.
The same idea in the exact infinite dimensional space is depicted in Figure

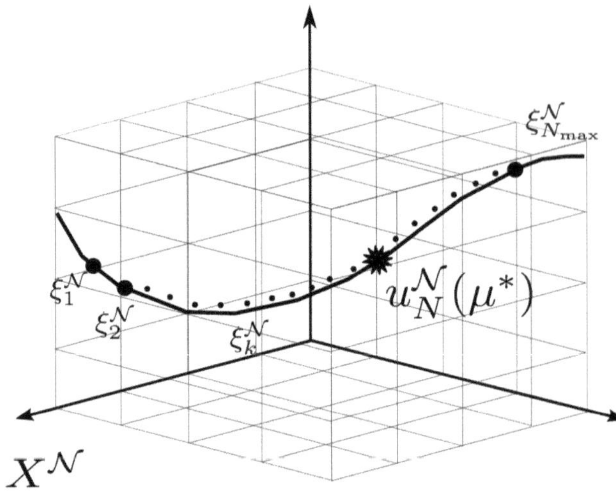

(a)

Figure 5: Approximation of $u_N^{\mathcal{N}}(\boldsymbol{\mu}^*)$

2.
It is thus wasteful to express the solution $u(\boldsymbol{\mu})$ as an arbitrary member of the unnecessarily rich space $X^{\mathcal{N}}$; rather, presuming that $\mathcal{M}$ is sufficiently smooth, we should represent $u(\boldsymbol{\mu})$ in terms of elements of an *ad-hoc* manifold much more lower dimensional (Patera and Rozza, 2007; Rozza et al., 2008).
The *RB* recipe is very simple (see Figure 5 for a graphical interpretation). The basic idea is to *efficiently* chose and compute N solutions or *"snapshots"* $\xi_1^{\mathcal{N}}, \xi_2^{\mathcal{N}}, \ldots, \xi_N^{\mathcal{N}} \in X^{\mathcal{N}}$ and then, for any arbitrary new $\boldsymbol{\mu}^* \in \mathcal{D}$, compute the solution $u_N^{\mathcal{N}}(\boldsymbol{\mu}^*)$ associated to this parameter thanks to an *appropriate combination* of the previously computed snapshots $\xi_k^{\mathcal{N}}$, $k = 1, \ldots, N$.
Note that "$u_N^{\mathcal{N}}(\boldsymbol{\mu})$" is not redundant; it means that this is the solution in the truth space $X^{\mathcal{N}}$ computed along the reduced manifold $\mathcal{M}^{\mathcal{N}}$, selecting N snapshots.
In the most part this work, if not specified, when dealing with *RB* solution we will always simply write $u_N(\boldsymbol{\mu})$ meaning the reduced solution in the truth space.
Now also the *Opportunity II* (section 1.2) can be understood; starting from the *RB* idea it is evident that are needed *at least* N solutions of the problem on the $\mathcal{N}$-dimensional *truth space.*
The *RB* approach is thus clearly *ill-suited* to the *single-query* or *few-query* situation; however, in the *real-time* and *many-query context* this *Offline* investment is readily acceptable in exchange for future asymptotic or Online

computational burden reduction.

3.2 RB spaces and bases

In this section the *RB* problem formulation in the *coercive* case is discussed.
We begin introducing the spaces and basis that allow us to build the reduced basis problem, subsequently the *creation* of the *RB* system, the *Offline/Online* procedure and the *a-posteriori* error bound will be introduced. There are different possible choices for the selection of the reduced basis spaces (Hermite, Lagrange, etc ...) that will lead ultimately to different reduced order model (Rozza et al., 2008; Patera and Rozza, 2007; Porsching, 1985; Ito and Ravindran, 2001). In the following the *Lagrange hierarchical spaces* used in this work will be discussed.

Lagrange hierarchical spaces. We introduce a set of linearly independent functions:

$$\xi^n \in X, \quad 1 \leq n \leq N_{\max} \tag{34}$$

where $N_{\max}$ is the maximum dimension of the *RB* space, in terms of which we define the *RB* approximation spaces:

$$X_N = \text{span}\{\xi^n, 1 \leq n \leq N\} \quad 1 \leq N \leq N_{\max} \tag{35}$$

where we assume, in order to build a "reduced basis" that the ξ^n are somehow related to the manifold $\mathcal{M}$.
By construction we obtain

$$X_N \subset X, \quad \dim(X_N) = N, \quad 1 \leq N \leq N_{\max} \tag{36}$$

moreover, as the same property holds recursively for any nested subset of X_N, we can say that the space X is *hierarchical.*

Definition 3.1. Given a space X, given $N_{\max}$ subsets of this space $X_n \subset X$, $1 \leq n \leq N_{\max}$, we say that X is a *hierarchical (or nested)* space if:

$$X_1 \subset X_2 \subset \ldots \subset X_{N_{\max}-1} \subset X_{N_{\max}}. \tag{37}$$

The hierarchical property (37), as we shall see, is important in ensuring (memory) efficiency for the resulting reduced basis approximation.
To introduce the Lagrange (hierarchical) *RB* recipe, we first define a master set of parameter points $\boldsymbol{\mu}^n \in \mathcal{D}$, $1 \leq n \leq N_{\max}$, we then define, for given $N \in \{1, \ldots, N_{\max}\}$, the Lagrange parameter samples

$$S_N = \{\boldsymbol{\mu}^1, \ldots, \boldsymbol{\mu}^N\}, \tag{38}$$

that we choose *nested* in order to build a hierarchical space, that is:

$$S_1 = \{\boldsymbol{\mu}^1\} \subset S_2 = \{\boldsymbol{\mu}^1, \boldsymbol{\mu}^2\} \subset \ldots \subset S_{N_{\max}}. \tag{39}$$

The associated *Lagrange RB* spaces are defined as:

$$W_N^{\mathcal{N}} = \text{span}\,\{u^{\mathcal{N}}(\boldsymbol{\mu}^n)\}, \quad 1 \le n \le N. \tag{40}$$

We observe that, by construction, these Lagrange spaces are hierarchical; in fact the samples S_N are nested thanks to the choice (39), then accordingly:

$$W_1^{\mathcal{N}} = \text{span}\left\{u^{\mathcal{N}}(\boldsymbol{\mu}^1)\right\} \subset W_2^{\mathcal{N}} = \text{span}\left\{u^{\mathcal{N}}(\boldsymbol{\mu}^2),\, u^{\mathcal{N}}(\boldsymbol{\mu}^1)\right\} \subset \ldots \subset W_{N_{\max}}^{\mathcal{N}} \tag{41}$$

The $u^{\mathcal{N}}(\boldsymbol{\mu}^n)$, $1 \le n \le N_{\max}$ are the so-called *"snaphots"* related to the low dimensional manifold $\mathcal{M}^{\mathcal{N}}$. As already mentioned in section 3.1, we would expect to well approximate any member of the manifold thanks to a good combination of the available snapshots.
In theory, in order to build the *RB* approximation spaces, it would be necessary to choose a set of parameter samples that induces a set of linearly independent snapshots; the *greedy* sampling, that will be introduced in section 3.3, induces linear dependent functions as N increases. In fact, if the snapshot chosen W_N contains much of the $\mathcal{D}$ induced manifold $\mathcal{M}$, then it will be clear that the new $\boldsymbol{\mu}^{N+1} \in \mathcal{D}$ will perforce be a combination of this functions.
We therefore pursue a Gram-Schmidt orthogonalization in the $(\cdot\,,\,\cdot)_X$ inner product to recover an orthonormal well-conditioned set of basis functions in order to guarantee a good algebraic stability without ill-conditioning (Patera and Rozza, 2007).

Orthogonal *RB* bases. To achieve the orthogonalization, we apply the already mentioned Gram-Schmidt standard orthogonalization (Meyer, 2000). Given the basis functions ξ^n, $1 \le n \le N_{\max}$ (34), that in the Lagrange space choice are the $u(\boldsymbol{\mu}_n)$, $1 \le n \le N_{\max}$ (40), we obtain the set of basis function ζ^n, $1 \le n \le N_{\max}$ as:

$$\begin{aligned}
&\zeta^1 = \xi^1/||\xi^1||_X;\\
&\texttt{for } n = 2 : N_{\max}\\
&\qquad z^n = \xi^n - \sum_{m=1}^{n-1}, (\xi^n, \zeta^m)_X\ \zeta^m;\\
&\qquad \zeta^n = z^n/||z^n||_X;\\
&\texttt{end}
\end{aligned} \tag{42}$$

As a result of this process we obtain the orthogonality condition:

$$(\zeta^m, \zeta^n)_X = \delta_{mn} \quad 1 \leq m, n \leq N_{\max} \tag{43}$$

where δ_{mn} is the Kronecker-delta symbol.
Finally we can express our reduced basis spaces X_N as:

$$X_N = \text{span}\{\zeta^n,\ 1 \leq n \leq N\} \quad 1 \leq N \leq N_{\max}. \tag{44}$$

Now any function $w_N \in X_N$ can be expressed as a linear combination of the reduced base X_N as:

$$w_N = \sum_{n=1}^{N} w_{N_n}\, \zeta^n \quad 1 \leq N \leq N_{\max} \tag{45}$$

for a unique combination of (RB) coefficients $w_{N_n} \in \mathbb{R}$, $1 \leq n \leq N_{\max}$.

Algebraic representation of *RB* bases. We now reconsider the orthogonalization process in order to introduce some concepts that will be necessary to build our *RB* problem starting from the *FE* original frame.
If we express our snapshots ξ^n in terms of *FE* functions ϕ_i, $1 \leq i \leq \mathcal{N}$:

$$\xi^n = \sum_{i=1}^{\mathcal{N}} \xi_i^n\, \phi_i, \quad 1 \leq n \leq N_{\max}, \tag{46}$$

similarly we may express our *RB* orthogonalized functions ζ^n as

$$\zeta^n = \sum_{i=1}^{\mathcal{N}} \zeta_i^n\, \phi_i, \quad 1 \leq n \leq N_{\max}. \tag{47}$$

Now, in the two cases above, we sort the *FE* coefficients in an array

$$\underline{\xi}^n \equiv \{\ \xi_1^n\, \xi_2^n\, \ldots\, \xi_{\mathcal{N}}^n\ \}^T \quad 1 \leq n \leq, N_{\max} \tag{48}$$

$$\underline{\zeta}^n \equiv \{\ \zeta_1^n\, \zeta_2^n\, \ldots\, \zeta_{\mathcal{N}}^n\ \}^T \quad 1 \leq n \leq N_{\max}. \tag{49}$$

We then introduce the algebraic representation $\underline{\underline{\mathbb{X}}}^{\mathcal{N}} \in \mathbb{R}^{\mathcal{N}\times\mathcal{N}}$ of the inner product $(\cdot, \cdot)_X$:

$$\underline{\underline{\mathbb{X}}}_{ij}^{\mathcal{N}} = (\phi^i, \phi^j)_{X^{\mathcal{N}}} \quad 1 \leq i, j \leq \mathcal{N}. \tag{50}$$

The orthogonalization process can be now formulated as:

$$\begin{aligned}
&\underline{\zeta}^1 = \underline{\xi}^1 / \sqrt{\underline{\xi}^{1^T} \underline{\underline{\mathbb{X}}}\, \underline{\xi}^1};\\
&\texttt{for } n = 2 : N_{\max}\\
&\qquad \underline{z}^n = \underline{\xi}^n - \sum_{m=1}^{n-1}, \left(\underline{\xi}^n\, \underline{\underline{\mathbb{X}}}\, \underline{\zeta}^m\right)\, \underline{\zeta}^m;\\
&\qquad \underline{\zeta}^n = \underline{z}^n / \sqrt{\underline{z}^{nT} \underline{\underline{\mathbb{X}}}\, \underline{z}^n};\\
&\texttt{end}
\end{aligned} \tag{51}$$

Finally it is useful to introduce the *"basis"* matrices $\underline{\underline{Z}} \in \mathbb{R}^{\mathcal{N}\times N}$, $1 \leq N \leq N_{\max}$:

$$Z_{N_{j\,n}} = \zeta_j^n \quad 1 \leq j \leq \mathcal{N},\ 1 \leq n \leq N. \tag{52}$$

This matrices is built in such a way that the nth-column of the matrix is formed by the vector of *FE* coefficients $\zeta_{N_j}^n$, $1 \leq j \leq \mathcal{N}$ associated to the nth *RB* function.

Galerkin projection. The projection strategy used in order to obtain the *RB* approximation is given by a Galerkin projection, which is arguably the best approach. We remark that the *RB* weak formulation has formally the same appearance as the *"exact"* weak formulation (see equations (17)-(18), section 2.3); in this case we properly replace the *FE* truth functional space with the *RB* approximation space; in the next Section we will show how to obtain the latter from the former by means of an algebraic procedure.
The problem states: given $\boldsymbol{\mu} \in \mathcal{D}$, find $u_N \left(\equiv u_N^{\mathcal{N}}\right) \in X_N \left(X_N^{\mathcal{N}}\right)$ such that

$$a(u_N(\boldsymbol{\mu}), v; \boldsymbol{\mu}) = f(v; \boldsymbol{\mu}) \quad \forall\, v \in X_N \tag{53}$$

we evaluate

$$s_N(\boldsymbol{\mu}) = f(u_N(\boldsymbol{\mu})). \tag{54}$$

From coercivity and continuity hypothesis on a and f, our conforming reduced basis $X_N^{\mathcal{N}} \subset X^{\mathcal{N}}$, and from our assumption of linear independence of snapshots, the problem (53)-(54) admits an unique solution (Quarteroni and Valli, 1997; Quarteroni, 2013). Thanks to the Galerkin projection, the optimality results subsequently discussed holds (Rozza et al., 2008; Patera and Rozza, 2007).

Proposition 3.2. *For any $\boldsymbol{\mu} \in \mathcal{D}$ and $u_N(\boldsymbol{\mu})$ and $s_N(\boldsymbol{\mu})$ satisfying (53)-(54):*

$$|||u^{\mathcal{N}} - u_N^{\mathcal{N}}|||_{\boldsymbol{\mu}} = \inf_{w_N \in X_N^{\mathcal{N}}} |||u^{\mathcal{N}} - w_N|||_{\boldsymbol{\mu}} \tag{55}$$

$$||u^{\mathcal{N}} - u_N^{\mathcal{N}}||_X \leq \sqrt{\frac{\gamma^e(\boldsymbol{\mu})}{\alpha^e(\boldsymbol{\mu})}} \inf_{w_N \in X_N^{\mathcal{N}}} ||u^{\mathcal{N}} - w_N||_X \tag{56}$$

as regard the output optimality results, in the compliant case *we obtain:*

$$\begin{aligned} s_N - s_N^{\mathcal{N}} &= |||u^{\mathcal{N}} - u_N^{\mathcal{N}}|||_{\boldsymbol{\mu}}^2 \\ &= \inf_{w_N \in X_N^{\mathcal{N}}} |||u^{\mathcal{N}} - w_N|||_{\boldsymbol{\mu}}^2 \end{aligned} \tag{57}$$

and furthermore

$$0 \leq s_N - s_N^{\mathcal{N}} \leq \gamma^e(\boldsymbol{\mu}) \inf_{w_N \in X_N^{\mathcal{N}}} ||u^{\mathcal{N}} - w_N||_X^2, \tag{58}$$

where $\gamma^e(\boldsymbol{\mu})$ and $\alpha^e(\boldsymbol{\mu})$ are the continuity and coercivity constants, respectively (defined in (6) and (5)) in the exact space. It will be shown that these optimality results will be used even after, replacing the exact constants with those evaluated with the reduced space approximation.
It is also necessary to remark from (57) that, in the *compliant case*, the error on the output is the *square* of the error of the field variable: we have the so-called *"square effect"*, this is crucial for the input/output accuracy and efficiency of the method.
Last but not least, $s_N(\boldsymbol{\mu})$ is a *lower bound* for $s^{\mathcal{N}}$, in fact: (i) $s_N(\boldsymbol{\mu}) = a(u_N(\boldsymbol{\mu}), u_N(\boldsymbol{\mu}); \boldsymbol{\mu})$ is a positive quantity, and (ii) the error in the output is the *square* of the error of the field variable (Rozza et al., 2008).

Offline-Online procedures: algebraic formulation. The algebraic formulation for the coercive problem will be explained. The crucial point that will be treated is the Online/Offline splitting procedure; this procedure will be equipped with an operation count to highlight the potential computational saving offered by the *RB* method.
In order to apply the standard variational procedure to obtain the algebraic formulation of the problem, we first expand $u_N(\boldsymbol{\mu})$:

$$u_N(\boldsymbol{\mu}) = \sum_{j=1}^{N} u_N^j(\boldsymbol{\mu}) \tag{59}$$

now, inserting the expansion (59) in the problem (53) and choosing $v = \zeta^i, \quad 1 \le i \le N$ as our *test function*, we obtain the set of linear algebraic equations

$$\sum_{j=1}^{N} a(\zeta^j, \zeta^i; \boldsymbol{\mu})\, u_N^j(\boldsymbol{\mu}) = f(\zeta^i; \boldsymbol{\mu}), \quad 1 \le i \le N \tag{60}$$

for the reduced basis coefficients $u_N^j(\boldsymbol{\mu})$ $1 \le i \le N$.
The output can then be expressed as

$$s_N(\boldsymbol{\mu}) = \sum_{j=1}^{N} u_N^j(\boldsymbol{\mu})\, f(\zeta^j; \boldsymbol{\mu}) \tag{61}$$

We now express these operations in matrix form; we first introduce the array of RB basis coefficients $u_N(\boldsymbol{\mu})$ as

$$\underline{u}_N \equiv [u_{N_1} u_{N_2} \dots u_{N_N}] \tag{62}$$

It follows from (60) that $\underline{u}_N \in \mathbb{R}^N$ satisfies

$$\underline{\underline{A}}_N(\boldsymbol{\mu})\, \underline{u}_N(\boldsymbol{\mu}) = \underline{F}_N(\boldsymbol{\mu}) \tag{63}$$

where $\underline{\underline{A}}_N(\boldsymbol{\mu}) \in \mathbb{R}^{N\times N}$ is the "stiffness matrix", and $\underline{F}_N(\boldsymbol{\mu}) \in \mathbb{R}^N$ is the "load/source" array. This quantity are in particular given by:

$$\underline{\underline{A}}_{N_{i,j}}(\boldsymbol{\mu}) = a(\zeta^j, \zeta^i; \boldsymbol{\mu}) \quad 1 \le i, j \le N \tag{64}$$

and

$$\underline{F}_{N_i}(\boldsymbol{\mu}) = f(\zeta^i; \boldsymbol{\mu}) \quad 1 \le i \le N. \tag{65}$$

Finally, the output can now be expressed as

$$\underline{s}_N(\boldsymbol{\mu}) = \underline{F}_N^T\, \underline{u}_N \tag{66}$$

It follows from our assumption of linear independence of the snapshots that the stiffness matrix is symmetric and positive definite.

Offline-Online. It is now possible, starting from the algebraic problem, to introduce the *Offline/Online* procedure.
The reduced basis system (63) is clearly of small size, in fact it is an $N \times N$ linear system that requires $\mathcal{O}(N^3)$ operation to solve it, plus $\mathcal{O}(N)$ operation to obtain the output from the equation (66).
By appealing to our previous assumption of affine parameter dependence

discussed in section 2.2, from (11) and (15), the *stiffness matrix* and *load/source vector* can be expressed, respectively, as

$$a(\zeta^m, \zeta^n; \boldsymbol{\mu}) = \sum_{q=1}^{Q_a} \theta_a^q(\boldsymbol{\mu})\, a(\zeta^m, \zeta^n) \quad 1 \le m, n \le N \tag{67}$$

and

$$f(\zeta^n; \boldsymbol{\mu}) = \sum_{q=1}^{Q_f} \theta_f^q(\boldsymbol{\mu})\, f(\zeta^n) \quad 1 \le n \le N. \tag{68}$$

The Offline-Online decomposition is now possible.

Offline: In the offline part we form:

1. the parameter independent matrices $\underline{\underline{\mathbb{A}}}_N^q \in \mathbb{R}^{N \times N}$

$$\underline{\underline{\mathbb{A}}}^q_{N_{m,n}} = a^q(\zeta^m, \zeta^n), \quad 1 \le m, n \le N, \ 1 \le q \le Q_a \tag{69}$$

2. the parameter independent vectors $\underline{\mathbb{F}}_N^q \in \mathbb{R}^N$

$$\underline{\mathbb{F}}^q_{N_n} = f^q(\zeta^n), \quad 1 \le n \le N, \ 1 \le q \le Q_f \tag{70}$$

This operations $\mathcal{N}$-dependent and hence very expensive are computed *once*.

Online: In the online stage we assemble, for any new $\boldsymbol{\mu} \in \mathcal{D}$:

1. the *RB* reduced stiffness matrix $\underline{\underline{A}}_N(\boldsymbol{\mu})$

$$\underline{\underline{A}}_N(\boldsymbol{\mu}) = \sum_{q=1}^{Qa} \theta_a^q(\boldsymbol{\mu})\, \underline{\underline{\mathbb{A}}}_N^q \tag{71}$$

2. the *RB* reduced load/source vector $\underline{F}_N(\boldsymbol{\mu})$

$$\underline{F}_N(\boldsymbol{\mu}) = \sum_{q=1}^{Q_f} \theta_f^q(\boldsymbol{\mu})\, \underline{\mathbb{F}}_N^q \tag{72}$$

the operation count is actually $\mathcal{N}$-independent and hence very inexpensive.

Link between FE and RB. Before a detailed discussion over the operation count, it is necessary to provide the link between the *FE* and the *RB* stiffness matrix and load/source vector; it is worth to remark that this operation will be completed once in the offline stage.

In particular it can be showed that the former are linked to the latter via the *snapshots* matrices $\underline{\underline{\mathbb{Z}}} \in \mathbb{R}^{\mathcal{N}\times N}$, $1 \le N \le N_{\max}$.
The stiffness matrix (64) that can be written as:

$$a(\zeta^m, \zeta^n; \boldsymbol{\mu}) = \sum_{j=1}^{\mathcal{N}} \sum_{i=1}^{\mathcal{N}} \zeta_j^m \, a(\phi^i, \phi^j; \boldsymbol{\mu}) \, \zeta_i^n \quad 1 \le m, n \le N_{\max} \tag{73}$$

thanks to the definition of the *basis matrix* $\underline{\underline{\mathbb{Z}}}$ (52) and the *FE* development of the reduced basis functions ζ^n (47) we may rewrite the stiffness matrix (73) as

$$\underline{\underline{A}}_N(\boldsymbol{\mu}) = \underline{\underline{\mathbb{Z}}}^T \underline{\underline{A}}(\boldsymbol{\mu}) \, \underline{\underline{\mathbb{Z}}}. \tag{74}$$

In the same way the reduced parametric independent stiffness matrices $\underline{\underline{A}}_N^q$ (71) are linked to the *FE* matrices:

$$\underline{\underline{\mathbb{A}}}_N^q = \underline{\underline{\mathbb{Z}}}^T \underline{\underline{\mathbb{A}}}^q \, \underline{\underline{\mathbb{Z}}}. \tag{75}$$

The load/source vectors (66) admit a similar treatment

$$\underline{F}_N(\boldsymbol{\mu}) = \underline{\underline{\mathbb{Z}}}^T \underline{F}(\boldsymbol{\mu}). \tag{76}$$

finally the parameter independent load/source vector 72 can be written as

$$\underline{\mathbb{F}}_N^q = \underline{\underline{\mathbb{Z}}}^T \underline{\mathbb{F}}^q(\boldsymbol{\mu}) \quad 1 \le q \le Q_f. \tag{77}$$

Operation count and storage. Thanks to the Offline/Online splitting procedure, we have achieved an Online $\mathcal{N}$-independent stage, hence very inexpensive.
It is necessary to focus on the Offline and Online complexity to quantify the computational reduction provided by the *RB* method.
We make use of Table 1 to summarize the computational burden requested to perform: (i) Offline, the *RB* basis assembling, (ii) Online, a single input/output evaluation.

In this table we have denoted with "*A*-matvec" and "$X^{\mathcal{N}}$-inprod", the matrix-vector multiplication and the inner product between two vectors $\in X^{\mathcal{N}}$, respectively.
As regard the storage we need only to store $\underline{\underline{\mathbb{A}}}_{N_{\max}}^q$ $1 \le q \le Q_a$ and $\underline{\mathbb{F}}_{N_{\max}}^q$ $1 \le q \le Q_f$, then extract only the sub-matrices and sub-vectors of desired N thanks to the hierarchical base property, as explained in (37).

3.3 Sample/space assembling

We now discuss the procedure used to select the snapshots in order to assemble the reduced basis approximation spaces, after a few preliminaries. We then turn to the *Greedy sampling* strategy exploited in this chapter

PART	ITEM	BURDEN	EQUATION
Offline	$\underline{\underline{\mathbb{Z}}}_N$	$N\,\mathcal{O}(\mathcal{N}^3)$	(52)
	$\underline{\underline{\mathbb{A}}}_N^q$	$Q_a\,N\,\underline{\underline{A}}$-matvec $+\,Q_a\,N^2\,X^{\mathcal{N}}$-inprod	(75)
	$\underline{\mathbb{F}}_N$	$Q_f\,N\,X^{\mathcal{N}}$-inprod	(77)
Online	$\underline{\underline{A}}_N$	$Q_a\,N^2$	(71)
	$\underline{F}_N$	$Q_f\,N$	(72)
	u_N	$\mathcal{O}(N^3)$	(60)
	s_N	N^2	(66)

Table 1: Offline/Online: coercive case

(Patera and Rozza, 2007). See also (Huynh et al., 2007; Rozza et al., 2008) for more options sampling strategies.
We shall denote by Ξ a finite sample of points in $\mathcal{D}$. These "test" samples Ξ serve as surrogates for $\mathcal{D}$ in the calculation and presentation of errors over the parameter domain. Typically these samples are chosen by *Monte Carlo* methods with respect to a uniform or log $-$uniform density.
Concerning the dimension of the sample, we always ensure that Ξ is sufficiently large that the reported results are *insensitive* to further refinement of the parameter sample.

Definition 3.3. Given a function $y : \mathcal{D} \to \mathbb{R}$, we define the $L^\infty(\Xi)$ and $L^p(\Xi)$ norms respectively as:

$$\begin{aligned} ||y||_{L^\infty(\Xi)} &\equiv \max_{\boldsymbol{\mu}\in\Xi} |y(\boldsymbol{\mu})| \\ ||y||_{L^p(\Xi)} &\equiv \left(|\Xi|^{-1} \sum_{\boldsymbol{\mu}\in\Xi} |y(\boldsymbol{\mu})|^p \right)^{1/p} . \end{aligned} \tag{78}$$

Definition 3.4. Given a function $z : \mathcal{D} \to X^{\mathcal{N}}$ (or X^e), we define the $L^\infty(\Xi; X)$ and $L^p(\Xi; X)$ norms respectively as:

$$\begin{aligned} ||z||_{L^\infty(\Xi;X)} &\equiv \max_{\boldsymbol{\mu}\in\Xi} |z(\boldsymbol{\mu})|_X \\ ||z||_{L^p(\Xi;X)} &\equiv \left(|\Xi|^{-1} \sum_{\boldsymbol{\mu}\in\Xi} ||z(\boldsymbol{\mu})||_X^p \right)^{1/p} . \end{aligned} \tag{79}$$

Here $|\Xi|$ denotes the cardinality of (the finite number of elements in) the test sample Ξ. We now introduce the *Greedy Lagrange spaces*, that will be used to build our *RB* approximation.

Greedy Lagrange spaces. We have already introduced the concept of *Lagrange spaces* (see section 3.2), we now have to extend this idea to Greedy Lagrange spaces.
We remark that this strategy is not indispensable to build a basic model reduction, but it is a *prerogative* of the *Reduced Basis method.* In fact the *Greedy* sampling, we are going to discuss, can be efficiently exploited only combined to an Offline/Online splitting procedure.
The idea of this strategy is starting with a train sample Ξ_{train}, we select N parameters $\boldsymbol{\mu}_1, \dots, \boldsymbol{\mu}_N$ and, as already seen in section 3.2, we form the reduced basis space X^N as:

$$X^N = \text{span}\left\{\xi_n = u^{\mathcal{N}}(\boldsymbol{\mu_n}),\ 1 \leq n \leq N\right\}. \tag{80}$$

More precisely, for the *Greedy approach*, we need a also *sharp*, *rigorous* and *efficient bound* $\Delta_N^{en}(\boldsymbol{\mu})$ for the reduced basis error $||u^{\mathcal{N}}(\boldsymbol{\mu}) - u_N(\boldsymbol{\mu})||_X$, where u_N is our *RB* approximation associated with the space X_N (Rozza et al., 2008). The superscript ^{en} denotes that the bound is related to the *energy norm* of the error, other options are discussed in Patera and Rozza (2007).
To quantify the *sharpness* and *rigour* properties, we recall the *effectivity* of an error bound.

Definition 3.5. The *effectivity* of an error bound, denoted by η, is defined as follows

$$\eta_N^{en} = \frac{\Delta_N^{en}}{||u^{\mathcal{N}}(\boldsymbol{\mu}) - u_N(\boldsymbol{\mu})||_X} \tag{81}$$

we require that

$$1 \leq \eta_N^{en} \leq \eta_{\max,\text{UB}}^{en} \quad \forall\, \boldsymbol{\mu} \in \mathcal{D},\ 1 \leq N \leq N_{\max} \tag{82}$$

where $\eta_{\max,\text{UB}}^{en}$ is finite and N independent.

It is possible to show that the inequality (82) is always fullfilled in the *RB* method (Rozza et al., 2008). The *rigour* property is illustrated by the *left* inequality: the error bound $\Delta_N^{en}(\boldsymbol{\mu})$ is never smaller than the true error $||u^{\mathcal{N}}(\boldsymbol{\mu}) - u_N(\boldsymbol{\mu})||_X$. The *sharpness* property is illustrated by the *right* inequality: $\Delta_N^{en}(\boldsymbol{\mu})$ *has not to bee much bigger* than the true error. Last, *efficient* means that the evaluation of $\Delta_N^{en}(\boldsymbol{\mu})$ is $\mathcal{N}$ independent, thanks to the Offline/Online procedure that we will show in section 3.4. The last property is crucial in the *Greedy* procedure, in fact it permits us to exploit a *very large* train sample Ξ_{train} in order to select the best snapshots to be include in our *RB* approximation spaces.

Greedy algorithm. We define $\overline{N}_{\max}$, an upper bound for $N_{\max}$ and $\epsilon_{\text{toll,min}}$ the desired minimum tolerance over the error bound.
Given Ξ_{train}, $S_1 = \{\boldsymbol{\mu}^1\}$ and $X^1 = \text{span}\{u^{\mathcal{N}}(\boldsymbol{\mu}^1)\}$,

$$
\begin{aligned}
&\texttt{for } N = 2 : \overline{N}_{\max} \\
&\qquad \boldsymbol{\mu}^N = \arg \max_{\boldsymbol{\mu}\in\Xi_{\text{train}}} \\
&\qquad \epsilon_{N-1} = \Delta_N^{en}(\boldsymbol{\mu}^N) \\
&\qquad \texttt{if } \epsilon_{N-1} \leq \epsilon_{\text{toll,min}} \\
&\qquad\qquad N_{\max} = N-1 \\
&\qquad \texttt{end} \\
&\qquad S_N = S_{N-1} \cup \boldsymbol{\mu}^N \\
&\qquad X_N = X_{N-1} + \text{span}\{u^{\mathcal{N}}(\boldsymbol{\mu}^N)\} \\
&\texttt{end}
\end{aligned}
\tag{83}
$$

In the *Greedy* algorithm the key point is to exploit an approximated (very cheap) error bound $\Delta_N^{en}(\boldsymbol{\mu}^N)$ instead of the true error (hence very expensive) $||u^{\mathcal{N}}(\boldsymbol{\mu}) - u_N(\boldsymbol{\mu})||$.
We remark that the *Greedy* algorithm *heuristically* minimizes the *RB* error bound in $L^\infty(\Xi_{\text{train}}; X)$ norm (Patera and Rozza, 2007; Rozza et al., 2008): the algorithm evaluates the error bounds $\forall\, \boldsymbol{\mu} \in \Xi_{\text{train}}$, then the next snapshot is selected such that it corresponds to the *maximum error bound.*

3.4 A-posteriori error bound

A-posteriori error bounds are crucial in the *RB* methodology. They are important for both *efficiency* and *reliability* of *RB* approximations.
As regards *efficiency*, error bounds play a role in *Offline* and *Online* stage. In the *Greedy* algorithm for example, the application of *error bounds* permits larger training sample at reduced *Offline* computational cost. Hence, we have a better accuracy of the reduced basis approximation which can be obtained with a smaller number N of basis functions, and hence we have a further reduction in the *Online* computational cost.
In other words, a posteriori error estimation permits us to control the error

thus allowing us to *minimize* the computational effort (Patera and Rozza, 2007).
As regards *reliability*, our *Offline* sampling procedures could not be exhaustive without a Greedy approach. For a large number of parameters P, there would be a large portion of the parameter space $\mathcal{D}$ which would remain unexplored. So, the error of a large parts of the parameter domain $\mathcal{D}$ would be affected by uncertainty.
The *a-posteriori* error bounds permit to *rigorously* bound the error for all new value of parameter $\boldsymbol{\mu}^* \in \mathcal{D}$. So we do not lose any confidence in the solution compared to the underlying *FE* solution while exploiting the rapid predictive power of the *RB* approximation.
As mentioned in section 3.3, the *a-posteriori* error bound must be rigorous (greater or equal to the true error) for all N and all parameters values in the parameter domain $\mathcal{D}$. Second, the bound must be reasonably *sharp*. An overly conservative error bound can yield inefficient approximations, typically N too large, or *suboptimal* engineering results, for example too much big safety margins.
For the coercive case, see Patera and Rozza (2007); Rozza et al. (2008), whereas for the non-coercive case see Rozza (2009a); Rovas (2003).

Preliminaries. We define the residual $r : \mathcal{D} \to \left(X^{\mathcal{N}}\right)'$ as

$$r(v;\boldsymbol{\mu}) = f(v;\boldsymbol{\mu}) - a(u_N^{\mathcal{N}}(\boldsymbol{\mu}, v;\boldsymbol{\mu})) \quad \forall\, v \in X^{\mathcal{N}} \tag{84}$$

where $\left(X^{\mathcal{N}}\right)'$ is the dual space of $X^{\mathcal{N}}$.
We also introduce the function $\hat{e} : \mathcal{D} \to X^{\mathcal{N}}$, the *Riesz* representation of $r(v;\boldsymbol{\mu})$, see (Quarteroni, 2013):

$$(\hat{e}(\boldsymbol{\mu}), v)_X = r\,(v;\boldsymbol{\mu}) \quad \forall\, v \in X^{\mathcal{N}}. \tag{85}$$

Finally, introducing the real error $e^{\mathcal{N}}(\boldsymbol{\mu})\ (\equiv e(\boldsymbol{\mu}))$

$$e(\boldsymbol{\mu}) = u^{\mathcal{N}}(\boldsymbol{\mu}) - u_N^{\mathcal{N}}. \tag{86}$$

Recalling that $u^{\mathcal{N}}(\boldsymbol{\mu})$ and $u_N(\boldsymbol{\mu})$ satisfies the equations (26) and (53), respectively, we get from (84), (85) that the error $e(\boldsymbol{\mu})$ satisfies the following relation

$$a(e(\boldsymbol{\mu}), v;\boldsymbol{\mu}) = r(v;\boldsymbol{\mu}) = (\hat{e}(\boldsymbol{\mu}), v) \quad \forall\, v \in X^{\mathcal{N}}. \tag{87}$$

We note that for our choice of inner product (25): $\hat{e}(\boldsymbol{\mu}) = e(\boldsymbol{\mu})$.
We then define the *dual norm* of $r(\,\cdot\,;\boldsymbol{\mu})$ associated to the dual space $\left(X^{\mathcal{N}}\right)'$:

$$||r(v;\boldsymbol{\mu})||_{X'} = \sup_{v\in X} \frac{r(v;\boldsymbol{\mu})}{||v||_X} = ||\hat{e}(\boldsymbol{\mu})||_X. \tag{88}$$

Note that the second equality follows from the *Riesz representation theorem.* This definition is crucial for the *Offline-Online* procedure.
Our aim is to find an approximated lower bound for $\alpha^{\mathcal{N}}(\boldsymbol{\mu})$ (5), that is a function $\alpha^{\mathcal{N}}_{\text{LB}} : \mathcal{D} \to \mathbb{R}$ such that

1. $0 < \alpha^{\mathcal{N}}_{\text{LB}}(\boldsymbol{\mu}) \leq \alpha^{\mathcal{N}}(\boldsymbol{\mu}) \quad \forall\, \boldsymbol{\mu} \in \mathcal{D}$;
2. the evaluation $\boldsymbol{\mu} \to \alpha^{\mathcal{N}}_{\text{LB}}$ should be independent of $\mathcal{N}$.

We will discuss the procedure to evaluate this coercivity lower bound in 3.5.

Error bound estimators. Now we can define our *energy*, *output* and *relative output error bound* estimators, that are defined respectively as (Patera and Rozza, 2007; Rozza et al., 2008):

$$\Delta_N^{en}(\boldsymbol{\mu}) = \frac{||\hat{e}(\boldsymbol{\mu})||_X}{(\alpha_{\text{LB}^{\mathcal{N}}}(\boldsymbol{\mu}))^{1/2}}, \tag{89a}$$

$$\Delta_N^{s}(\boldsymbol{\mu}) = \frac{||\hat{e}(\boldsymbol{\mu})||_X^2}{\alpha_{\text{LB}^{\mathcal{N}}}(\boldsymbol{\mu})}, \tag{89b}$$

$$\Delta_N^{s,rel}(\boldsymbol{\mu}) = \frac{||\hat{e}(\boldsymbol{\mu})||_X^2}{\alpha_{\text{LB}^{\mathcal{N}}}(\boldsymbol{\mu})\, s_N^{\mathcal{N}}(\boldsymbol{\mu})} = \frac{\Delta_N^{s}(\boldsymbol{\mu})}{s_N^{\mathcal{N}}(\boldsymbol{\mu})}. \tag{89c}$$

Effectivity estimators. As already discussed in Section 3.3, associated to each estimator theres is an *effectivity estimator* as a measure of the *quality* of the error bound estimators and are needed to certify that the *RB* method is *rigorous* and *sharp*. We introduce the following ones:

$$\eta_N^{en}(\boldsymbol{\mu}) = \frac{\Delta_N^{en}(\boldsymbol{\mu})}{|||e(\boldsymbol{\mu})|||_{\boldsymbol{\mu}}}, \tag{90a}$$

$$\eta_N^{s}(\boldsymbol{\mu}) = \frac{\Delta_N^{s}(\boldsymbol{\mu})}{s^{\mathcal{N}}(\boldsymbol{\mu}) - s_N^{\mathcal{N}}(\boldsymbol{\mu})}, \tag{90b}$$

$$\eta_N^{en,rel}(\boldsymbol{\mu}) = \frac{\Delta_N^{s,rel}(\boldsymbol{\mu})}{\left(s^{\mathcal{N}}(\boldsymbol{\mu}) - s_N^{\mathcal{N}}(\boldsymbol{\mu})\right)/s^{\mathcal{N}}(\boldsymbol{\mu})}. \tag{90c}$$

It is can be shown that the *effectivities* are a measure of the *rigor* and *sharpness* for an error bound.

Proposition 3.6. *The following results holds (see Rozza et al. (2008) for the proof):*

$$1 < \eta_N^{en}(\boldsymbol{\mu}) \leq \sqrt{\frac{\gamma^e(\boldsymbol{\mu})}{\alpha^{\mathcal{N}}_{\text{LB}}(\boldsymbol{\mu})}}, \tag{91}$$

$$1 < \eta_N^{s}(\boldsymbol{\mu}) \leq \frac{\gamma^e(\boldsymbol{\mu})}{\alpha^{\mathcal{N}}_{\text{LB}}(\boldsymbol{\mu})}, \tag{92}$$

and finally, with regard to $\eta_N^{en,rel}(\boldsymbol{\mu})$*, it can be shown that:*

$$\eta_N^{en,rel}(\boldsymbol{\mu}) = (\eta_N^{en}(\boldsymbol{\mu}))^2. \tag{93}$$

Offline-Online procedure. The main component of the error bound is the computation of the dual norm of the residual $||\hat{e}(\boldsymbol{\mu})||_X$. To develop the Offline-Online procedure, we introduce the residual expansion, $\forall v \in X$:

$$r(v;\boldsymbol{\mu}) = \sum_{q=1}^{Q_f} \theta_f^q(\boldsymbol{\mu})\, f^q(v) + \sum_{q=1}^{Q_a}\sum_{n=1}^{N} \theta_a^q(\boldsymbol{\mu})\, u_{N_n}(\boldsymbol{\mu})\, a^q\,(\xi_n, v)\,. \tag{94}$$

This expansion directly follows from our affine assumption (11) and from the *RB* development $u_N(\boldsymbol{\mu}) = \sum_{n=1}^{N} u_{N_n}\,\xi_n$.
Moreover, we have from (87) that:

$$(\hat{e}(\boldsymbol{\mu}), v)_X = \sum_{q=1}^{Q_f} \theta_f^q(\boldsymbol{\mu})\, f^q(v) + \sum_{q=1}^{Q_a}\sum_{n=1}^{N} \theta_a^q(\boldsymbol{\mu})\, u_{N_n}(\boldsymbol{\mu})\, a^q\,(\xi_n, v)\,. \tag{95}$$

Consequently, defining

$$(\mathcal{C}^q, v)_X = f^q(v) \qquad\qquad 1 \le q \le Q_f \tag{96a}$$
$$(\mathcal{L}_n^q, v)_X = -a^q(\xi_n, v) \qquad\qquad 1 \le q \le Q_a,\ 1 \le n \le N \tag{96b}$$

we can write

$$\hat{e}(\boldsymbol{\mu}) = \sum_{q=1}^{Q_f} \theta_f^q(\boldsymbol{\mu})\, \mathcal{C}^q + \sum_{q=1}^{Q_a} u_{N_n}(\boldsymbol{\mu})\, \theta_a^q(\boldsymbol{\mu})\, \mathcal{L}_n^q. \tag{97}$$

We remark that (96a) and (96b) are parameter-independent *Poisson-like* problems, hence $\mathcal{C}^q$ and $\mathcal{L}_n^q$ are computed *Offline*. We thus obtain

$$
\begin{aligned}
||\hat{e}(\boldsymbol{\mu})||_X &= \left(\sum_{q=1}^{Q_f} \theta_f^q(\boldsymbol{\mu})\, \mathcal{C}^q + \sum_{q=1}^{Q_a} \sum_{n=1}^{N} \theta_a^q(\boldsymbol{\mu})\, u_{n_N}(\boldsymbol{\mu})\, \mathcal{L}_n^q, \right)_X \\
&= \sum_{q=1}^{Q_f} \sum_{q'=1}^{Q_f} \theta_f^q(\boldsymbol{\mu}) \theta_f^{q'}(\boldsymbol{\mu}) \left(\mathcal{C}^q, \mathcal{C}^{q'} \right)_X + \\
&\quad + \sum_{q=1}^{Q_a} \sum_{n=1}^{N} \theta_a^q(\boldsymbol{\mu})\, u_{N_n} \left\{ 2 \sum_{q'=1}^{Q_f} \theta_a^q(\boldsymbol{\mu}) \theta_f^{q'}(\boldsymbol{\mu}) \left(\mathcal{L}_n^q, \mathcal{C}^{q'} \right)_X + \right. \\
&\quad \left. + \sum_{q'=1}^{Q_a} \sum_{n'=1}^{N} \theta_a^{q'} \left(\mathcal{L}_n^q, \mathcal{L}_{n'}^{q'} \right)_X \right\}.
\end{aligned} \tag{98}
$$

The Offline-Online procedure is clear. In the Offline stage, we first compute $\mathcal{C}^q$, $1 \leq q \leq Q_f$ and $\mathcal{L}_n^q$, $1 \leq q \leq Q_a$, $1 \leq n \leq N$, then we compute and store the quantities:

$$\left(\mathcal{C}^q, \mathcal{C}^{q'} \right)_X \qquad 1 \leq q \leq Q_f,\ 1 \leq q' \leq Q_f \tag{99}$$

$$(\mathcal{L}_n^q, \mathcal{C}^q)_X \qquad 1 \leq q \leq Q_a,\ 1 \leq q' \leq Q_f \tag{100}$$

$$(\mathcal{L}_n^q, \mathcal{L}^q)_X \qquad 1 \leq q \leq Q_a,\ 1 \leq q \leq Q_a \tag{101}$$

In the *Online* stage we evaluate the expression 98 which consists in a sum. The computational cost to perform this evaluation is:

$$n^2 Q_a^2 + 2nQ_aQ_f + nQ_f^2, \tag{102}$$

so it is $\mathcal{N}$ independent, hence very inexpensive.

3.5 Successive constraint method

We now discuss the successive constraint method (*SCM*). This tool enables the construction of lower (and upper) bounds for the *coercivity* (and *inf-sup* stability) constants (5), required in a posteriori error analysis of *RB* approximations: Without risk of a global comprehension loss, the reader can go to proposition 3.7. The method, based on an *Offline-Online* strategy, reduces the *Online* calculation to a small *Linear Programming problem*: the *objective* is a parametric expansion of the underlying *Rayleigh* quotients, the *constraints* reflect stability information at optimally selected parameter points. The *state of the art* method is presented in Huynh et al. (2007), see also Rozza et al. (2008).

Coercive case. We define

$$\mathcal{Y} \equiv \left\{ y = (y_1 \ldots y_{Q_a}) \in \mathbb{R}^{Q_a} \mid \exists w_y \in X^{\mathcal{N}} \text{ s.t. } y_q = \frac{a^q(w_y, w_y)}{||w||_{X^{\mathcal{N}}}}, \ 1 \le q \le Q_a \right\}. \tag{103}$$

We further define the *objective function* $\mathscr{F} : \mathcal{D} \times \mathbb{R}^{Q_a} \to \mathbb{R}$ as

$$\mathscr{F}(y; \boldsymbol{\mu}) = \sum_{q=1}^{Q_a} \theta_a^q(\boldsymbol{\mu})\, y_q. \tag{104}$$

We may then write our *coercivity* constant as

$$\alpha^{\mathcal{N}}(\boldsymbol{\mu}) = \min_{y \in \mathcal{Y}} \mathscr{F}(y; \boldsymbol{\mu}). \tag{105}$$

We next introduce a *constraint box* that is the set of all the feasible value for y, defined as

$$\begin{aligned} \mathcal{B} &= \prod_{q=1}^{Q_a} \{\sigma_-^q, \sigma_+^q\} \\ &= \prod_{q=1}^{Q_a} \left\{ \inf_{w \in X^{\mathcal{N}}} \frac{a^q(w,w)}{||w||^2_{X^{\mathcal{N}}}}, \ \sup_{w \in X^{\mathcal{N}}} \frac{a^q(w,w)}{||w||^2_{X^{\mathcal{N}}}} \right\}. \end{aligned} \tag{106}$$

We also introduce the two parameter sets $\mathcal{S}$ and $\mathcal{P}$, that will be used to define the stability and positivity constraint, respectively:

$$\mathcal{S} = \{s_1 \in \mathcal{D}, \ldots, s_k \in \mathcal{D}\}, \tag{107}$$

$$\mathcal{P} = \{p_1 \in \mathcal{D}, \ldots, p_k \in \mathcal{D}\}. \tag{108}$$

Moreover, for any finite-dimensional *subset* of $\mathcal{D}$ ($= \mathcal{S}$ or $\mathcal{P}$), we denote with $\mathcal{S}^{\mathrm{M},\boldsymbol{\mu}}$ (or $\mathcal{P}^{\mathrm{M},\boldsymbol{\mu}}$) the set of M points closest to[4] $\boldsymbol{\mu}$ in $\mathcal{S}$ (or $\mathcal{P}$).
If $\mathrm{M} > |\mathcal{S}|$, (or $> |\mathcal{P}|$)[5], then $\mathcal{S}^{\mathrm{M},\boldsymbol{\mu}} = |\mathcal{S}|$ (or $\mathcal{P}^{\mathrm{M},\boldsymbol{\mu}} = \mathcal{P}$).

[4] In the *Euclidean norm*

[5] We recall that $|\cdot|$ denotes the *cardinality* of a finite set of elements

Lower and Upper bound. For given $\mathcal{S} \subset \mathcal{D}$, $\mathrm{M}_\alpha \in \mathbb{N}$ (stability constraints), $\mathrm{M}_+ \in \mathbb{N}$ (positivity constraints), we define the *lower bound* set as

$$\mathcal{Y}_{\mathrm{LB}}(\mathcal{S};\boldsymbol{\mu}) \equiv \left\{ y \in \mathcal{B} \mid \sum_{q=1}^{Q_a} \theta_a^q(\boldsymbol{\mu}') y_q \geq \alpha^{\mathcal{N}}(\boldsymbol{\mu}'),\ \forall \boldsymbol{\mu}' \in \mathcal{S}^{\mathrm{M}_\alpha,\boldsymbol{\mu}}; \right.$$
$$\left. \sum_{q=1}^{Q_a} \theta_a^q(\boldsymbol{\mu}') y_q \geq 0,\ \forall \boldsymbol{\mu}' \in \mathcal{P}^{\mathrm{M}_+,\boldsymbol{\mu}} \right\}. \tag{109}$$

Furthermore we define the *upper bound* set as

$$\mathcal{Y}_{\mathrm{UB}}(\mathcal{S}) \equiv \{y^\star(\boldsymbol{\mu})(s_k),\ 1 \leq k \leq |\mathcal{S}|\} \tag{110}$$

for

$$y^\star(\boldsymbol{\mu}) \equiv \arg\min_{y \in \mathcal{Y}} \mathscr{F}(y;\boldsymbol{\mu}). \tag{111}$$

Finally we obtain the *coercivity lower* and *upper bound* as

$$\alpha_{\mathrm{LB}}(\mathcal{S};\boldsymbol{\mu}) = \min_{y \in \mathcal{Y}_{\mathrm{LB}}(y;\mathcal{S})} \mathscr{F}(y;\boldsymbol{\mu}), \tag{112}$$

$$\alpha_{\mathrm{UB}}(\mathcal{S};\boldsymbol{\mu}) = \min_{y \in \mathcal{Y}_{\mathrm{UB}}(\mathcal{S})} \mathscr{F}(y;\boldsymbol{\mu}). \tag{113}$$

It is possible to show that the *lower/upper bounds* provided above, *effectively bound* the coercivity constant, the subsequent result holds:

Proposition 3.7. *Given* $\mathcal{S}$, $\mathcal{P}$ *and* $\mathrm{M}_\alpha \in \mathbb{N}$, $\mathrm{M}_+ \in \mathbb{N}$

$$\alpha_{\mathrm{LB}}(\mathcal{S};\boldsymbol{\mu}) \leq \alpha^{\mathcal{N}}(\boldsymbol{\mu}) \leq \alpha_{\mathrm{UB}}(\mathcal{S};\boldsymbol{\mu}) \qquad \forall \boldsymbol{\mu} \in \mathcal{D} \tag{114}$$

The proof can be found in Huynh et al. (2007); Rozza et al. (2008).

We expect that if $\mathcal{S}$ is sufficiently large, then

1. $y^\star(\boldsymbol{\mu})$ will be sufficiently close to a member of $\mathcal{Y}_{\mathrm{UB}}$ to provide a good *upper bound*;
2. the *stability* and *positivity* constraints in $\mathcal{Y}_{\mathrm{LB}}$ will sufficiently restrict y to provide a good *lower bound.*

SCM algorithm. We now present the algorithm to exploit the evaluation of the *coercivity* constant.
The task of the *SCM* is – given a sample train $\Xi_{\mathrm{SCM}} = \{\boldsymbol{\mu}^1_{\mathrm{SCM}}, \ldots, \boldsymbol{\mu}^{n_{\mathrm{SCM}}}_{\mathrm{SCM}}\}$ of dimension $|\Xi_{\mathrm{SCM}}| = n_{\mathrm{SCM}}$ – to select Greedy parameters in Ξ_{SCM} and

construct the sets $\mathcal{S}_k = \{s_1 = \boldsymbol{\mu}^1_{\text{SCM}} \cup \ldots \cup s_{K_{\max}} = \boldsymbol{\mu}^{K_{\max}}_{\text{SCM}}\}$. We now introduce the algorithm.
We define M_α, M_+, $\mathcal{P}$ and a tolerance $\epsilon_{\text{SCM}} \in \,]0,1[$, then we set $K_{\mathcal{S}} = 1$ and choose $\mathcal{S}_1 = \{s_1 = \boldsymbol{\mu}^1_{\text{SCM}}\}$ *arbitrarily*, then

$$
\begin{aligned}
&\texttt{while} \max_{\boldsymbol{\mu}\in\Xi_{\text{SCM}}} \left[\frac{\alpha_{\text{UB}}(\mathcal{S};\boldsymbol{\mu}) - \alpha_{\text{LB}}(\mathcal{S};\boldsymbol{\mu})}{\alpha_{\text{UB}}(\mathcal{S};\boldsymbol{\mu})}\right] > \epsilon_{\text{SCM}} \\
&\qquad s_{K+1} = \arg\max_{\boldsymbol{\mu}\in\Xi_{\text{SCM}}} \left[\frac{\alpha_{\text{UB}}(\mathcal{S};\boldsymbol{\mu}) - \alpha_{\text{LB}}(\mathcal{S};\boldsymbol{\mu})}{\alpha_{\text{UB}}(\mathcal{S};\boldsymbol{\mu})}\right] \\
&\qquad \mathcal{S}_{\mathcal{K}+\infty} = \mathcal{S}_{\mathcal{K}} \cup s_{K+1} \qquad\qquad (115)\\
&\qquad K = K+1 \\
&\texttt{end} \\
&K_{\max} = K
\end{aligned}
$$

Normally we set $\epsilon_{\text{SCM}} \approx 0.75$ which is a crude lower bound but with a little effect on error bounds, (Huynh et al., 2007).

Offline-Online procedure. We note that to compute the arg max we must solve a *linear optimization problem* or *Linear Program* (*LP*), for the lower bound $\alpha_{\text{LB}}(\boldsymbol{\mu})$ (112).
In the coercive case, the lower bound *LP*'s contains:

- design variables
  1. Q_a variables, $y = \{y_1, \ldots, y_{Q_a}\}$;
- constraints
  1. $2Q_a$ bounding boxes for $y \in \mathcal{B}$;
  2. M_α stability constraints;
  3. M_+ positivity constraints.

It is clear that the operation count for the *Online* stage $\boldsymbol{\mu} \to \alpha_{\text{LB}}(\boldsymbol{\mu})$ is *independent* of $\mathcal{N}$.
Nonetheless we first must determine our set $\mathcal{S}$ and obtain the $\alpha^{\mathcal{N}}(s_k)$, $1 \le k \le |\mathcal{S}|\,(\equiv K_{\mathcal{S}})$, by an Offline *Greedy SCM* algorithm.

Offline: In the *Offline* stage, we have to construct the set $\mathcal{B}$ (once) and then:

1. evaluate $\alpha^{\mathcal{N}}(s_k)$;
2. evaluate $y^\star(s_k)$;
3. form $\mathcal{Y}_{\text{LB}}$;
4. perform a lower bound *LP*'s to evaluate $\alpha_{\text{LB}}(s_k)$.

The first three quantities of course depends on $\mathcal{N}$, nonetheless it is important to *remark* that there are no *cross terms* $\mathcal{O}\,(n_{\text{SCM}} \times \mathcal{N})$.

Online: In the *Online* stage, given a new value $\boldsymbol{\mu}$ we have to perform a lower bound *LP*'s (*LP*) to evaluate $\alpha_{\mathrm{LB}}(\boldsymbol{\mu})$. This Online stage is hence independent on $\mathcal{N}$.
In the table 2 we summarize the computational cost to evaluate the *Offline/Online* stage of the *SCM*:

PART	ITEM	COMPLEXITY	EQUATION
Offline	$\mathcal{B}$	$2Q_a$-eigenproblems over $X^{\mathcal{N}}$	(106)
	$\alpha^{\mathcal{N}}(s_k)$	$K_{\max}$-eigenproblems over $X^{\mathcal{N}}$	(109)
	$y^{\star}(s_k)$	$K_{\max}\, Q_a$-inner product	(111)
	$\mathcal{Y}_{\mathrm{LB}}$	$\mathcal{N}\, Q_a\, K_{\max}$	(109)
	$\alpha_{\mathrm{LB}}(s_k)$	$n_{\mathrm{SCM}}\, K_{\max}$ *LP*'s of "size" $\mathcal{O}\,(2Q_a + \mathrm{M}_\alpha + \mathrm{M}_+)$	(112)
Online	$\alpha_{\mathrm{LB}}(\boldsymbol{\mu})$	1 *LP*'s of "size" $\mathcal{O}\,(2Q_a + \mathrm{M}_\alpha + \mathrm{M}_+)$	(72)

Table 2: Offline/Online: SCM

3.6 Choice of the truth approximation

It would be preferable to build the *RB* approach directly upon the exact solution, but this is not in general possible. As indicated earlier, the RB approximation shall be built upon and reduced basis error will be measured relative to a "truth" Galerkin *FE* approximation. Therefore it is necessary to choose *properly* the underlying discretization.

Choice of $\mathcal{N}$. In order to obtain a satisfying reduced basis model able to describe in an accurate way the exact behavior of the physical process, it is necessary to chose the discretization properly; that is in order to minimize the underlying error between *exact* solution and the *truth approximation*. Let $u^{RB}(\boldsymbol{\mu})$ be the *RB* solution of the problem and $u^{\mathcal{N}}(\boldsymbol{\mu})$ the finite element solution, than the error is the sum of (at least) two terms:

$$||u^e(\boldsymbol{\mu}) - u^{RB}(\boldsymbol{\mu})|| = \underbrace{||u^e(\boldsymbol{\mu}) - u^{\mathcal{N}}(\boldsymbol{\mu})||}_{\text{neglected}} + \underbrace{||u^{\mathcal{N}}(\boldsymbol{\mu}) - u_N(\boldsymbol{\mu})||}_{\text{considered}}. \qquad (116)$$

The minimization of the second addendum is a task delagated to the reduced basis method, on the contrary the minimization of the first is not related to the method.

Because of this, it is necessary to provide a feasible "starting point". This can be achieved thanks to the choice of a discretization method able to describe correctly the problem.
We shall require that our family of truth subspaces $X^{\mathcal{N}}$ satisfies the approximation condition:

$$\max_{\boldsymbol{\mu}\in\mathcal{D}} \inf_{w\in X^{\mathcal{N}}} ||u(\boldsymbol{\mu}) - w||_{X^e} \to 0 \quad \text{as} \quad \mathcal{M} \to \infty. \tag{117}$$

The choice of a *finite element* approximation automatically fulfill this requirement because the method is *strongly consistent* (Quarteroni and Valli, 1997; Quarteroni, 2013); thus for sufficiently large $\mathcal{N}$, it is possible to approximate $u^e(\boldsymbol{\mu})$ and $s^e(\boldsymbol{\mu})$ arbitrarily closely.
In particular we define the difference $\epsilon^{\mathcal{N}}$ between the exact solution and the approximation as:

$$\epsilon^{\mathcal{N}} = \max_{\boldsymbol{\mu}\in\mathcal{D}} ||u(\boldsymbol{\mu}) - u^{\mathcal{N}}(\boldsymbol{\mu})||_{X^e} \overset{\mathcal{N}\to\infty}{\longrightarrow} 0. \tag{118}$$

In general, $\mathcal{N}$ must be chosen rather large to achieve a reasonable engineering accuracy $\epsilon^{\mathcal{N}}$.
In 3D problems the complexity is higher since there is greater variability of the solution field as the parameters changes. Therefore it is necessary to discretize the problem so that for any possible combination of the parameters the accuracy is kept under a safe tolerance. In fact it is worth to recall that the *RB Offline* representation has to be built over a *unique truth representation* for all $\boldsymbol{\mu} \in \mathcal{D}$; the truth approximation is *"frozen"* in the *RB* methodology. The choice of the optimum (or at least of a suitable) *truth* solution is *not trivial*, nevertheless it is possible to verify *a-posteriori* if the discretization is enough rich to seize all the geometrical and physical complexity of the problem.

4 Geometrical parametrization

In this section we will introduce how to handle a domain which is parameter dependent.
The *RB* method described in section 3 requires that Ω is parameter independent: if we wish to consider linear combinations of snapshots, these snapshots must be defined relative to a common spatial configuration (domain) .
Then to allow geometrical variations, we must express Ω, our parameter independent domain, as the pre-image of Ω_o, the original (actual, deformed)

parameter dependent domain (Rozza et al., 2008).
The geometrical transformation will yield *variable* (parameter-dependent) *coefficients* in the reference-domain linear and bilinear forms that, under suitable hypotheses to be discussed below, will take at the end the requisite affine form (11).

4.1 Affine parametric precondition

We now introduce a domain decomposition:

$$\Omega_o(\boldsymbol{\mu}) = \bigcup_{k=1}^{K_{\text{dom}}} \Omega_o^k(\boldsymbol{\mu}) \tag{119}$$

where the $\Omega_o^k(\boldsymbol{\mu})$ are mutually non overlapping subdomains, that is for any $\boldsymbol{\mu} \in \mathcal{D}$

$$\Omega_o^k(\boldsymbol{\mu}) \cap \Omega_o^{k'}(\boldsymbol{\mu}) = 0 \quad 1 \leq k, k' \leq K_{\text{dom}},\ k \neq k'.$$

This coarse domain decomposition will be denoted *RB* discretization.
We now choose a parameter of reference $\boldsymbol{\mu}_{\text{ref}} \in \mathcal{D}$ and define our reference domain as $\Omega_r \equiv \Omega(\boldsymbol{\mu}_{\text{ref}})$.
We will never omit the pedix script beside the domain Ω we are dealing with, to avoid any confusion between the parameter dependent *original domain* $\Omega_o(\boldsymbol{\mu})$ (sometimes for brevity, just Ω_o) and the parameter independent reference domain Ω_r.
We will build our *FE* approximation on a very fine *FE* subtriangulation of the coarse *RB* decomposition.
This *FE* subtriangulation ensures that the *FE* approximation accurately treats the perhaps discontinuous coefficients (that could arise from property and geometry variation) associated with the different subdomains. The subtriangulation also plays an *important role* in the generation of the affine representation.
The choice of $\boldsymbol{\mu}_{\text{ref}}$ has to be done in an optimal way to reduce both Offline and Online computational effort.
Tipically the reference domain shall be built choosing a $\boldsymbol{\mu}_{\text{ref}}$ at the *"center"* of our parameter domain $\mathcal{D}$ in order to *minimize* the *distorsion* and consequently reduce the requisite $\mathcal{N}$.

We now state our *Affine Geometry Precondition.* We can treat any *original* domain $\Omega_o(\boldsymbol{\mu})$, that *admits* a *domain decomposition* (119), for which, $\forall\, \boldsymbol{\mu} \in \mathcal{D}$

$$\Omega_r = \mathcal{T}^{\text{aff},k}(\Omega o(\boldsymbol{\mu})^k; \boldsymbol{\mu}) \tag{120}$$

for *affine mappings* $\mathcal{T}^{\text{aff},k}(\,\cdot\,;\boldsymbol{\mu}) : \Omega_o(\boldsymbol{\mu}) \to \Omega_r$, $1 \le k \le K_{\text{dom}}$, that satisfy two requisites:

1. *individually bijective*;
2. *collectively continuos* (interface condition), that is, given two different subdomains denoted with k and k', $\forall \mathbf{x}_o \in \Omega_o^k(\boldsymbol{\mu}) \cap \Omega_o^{k'}$, holds the following condition

$$\mathcal{T}^{\text{aff},k}(\mathbf{x_o};\boldsymbol{\mu}) = \mathcal{T}^{\text{aff},k'}(\mathbf{x_o};\boldsymbol{\mu}). \tag{121}$$

We have depicted the idea of the affine transformation in Figure 6. Of

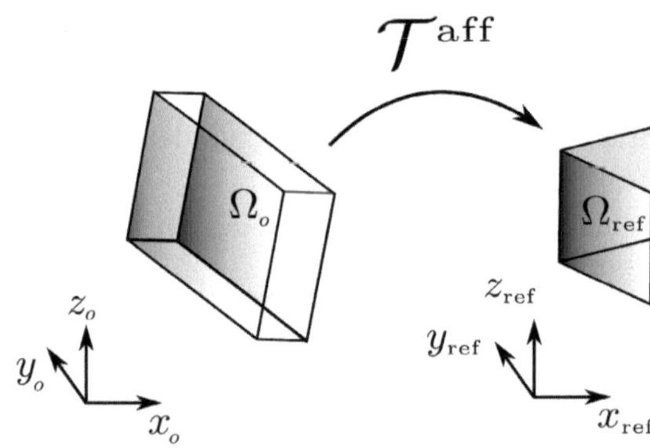

Figure 6: A 3D affine transformation

course, thanks to the requested bijective property, we can replace this definition with the forward version taking the inverse of $\mathcal{T}^{\text{aff}}$ committing any crime.

The *Affine Geometry Precondition* is a *necessary* condition for affine parameter dependence as defined in (11).

Note that we purposely define K_{dom} with respect to the *exact* problem, rather than the *FE* approximation: K_{dom} is not depending on $\mathcal{N}$.

We now give a more explicit representation of the *affine transformation* $\mathcal{T}^{\text{aff}}$, to better understand how this geometry precondition will be exploited.

We state that, for $1 \le k \le K_{\text{dom}}$, for any $\boldsymbol{\mu} \in \mathcal{D}$ and for all $\mathbf{x}_o \in \Omega_o^k(\boldsymbol{\mu})$, the affine transformation is defined as follows:

$$\mathbf{x}_r = \mathcal{T}_i^{\text{aff},k}(\mathbf{x}_o;\boldsymbol{\mu}) = C_i^{\text{aff},k}(\boldsymbol{\mu}) + \sum_{j=1}^{d} G_{ij}^{\text{aff},k}(\boldsymbol{\mu})\,\mathbf{x}_o, \quad 1 \le i \le d \tag{122}$$

for given $C^{\text{aff},k}(\boldsymbol{\mu}) : \mathcal{D} \to \mathbb{R}^d$ and $G^{\text{aff},k}(\boldsymbol{\mu}) : \mathcal{D} \to \mathbb{R}^{d\times d}$, that are called the *affine mapping coefficients*; we recall that d is the spatial dimension of the

problem, hence in our case $d = 3$.
The affine transformation is thus the superposition of a translation $C^{\text{aff}}(\boldsymbol{\mu})$, that do not modify the shape of the domain, and a deformation $G^{\text{aff}}(\boldsymbol{\mu})$ that can be a dilation/contraction or a shear. It is worth to remark that, in this work, the transformation *must* depends only upon the parameter $\boldsymbol{\mu}$.

A more general transformation, that involves a spatial coordinates dependence, is not considered in the framework of this thesis. This kind of transformation called *"nonaffine"* has been recently adopted in the context of the *RB* methodology (M. et al., 2004; Rozza, 2009a). The *non-affine* representation of the geometry arises from the so-called *free form deformation* techniques, which are very well suited, for example, for shape optimizations of complex geometries (Manzoni et al., 2012; Lassila and Rozza, 2010) and (Rozza and Manzoni, 2010).
The basic idea is that, thanks to an highly specialized technique, the so-called *empirical interpolation* (*EIM*), it is possible to approximate a *non-affine* tranformation thanks to a *superposition* of different *affine* transformations.
Nonetheless this results has been established in a 2D context, the extension to the 3D case in still under investigation.

We can now define the associated *Jacobians*:

$$J^{\text{aff},k}(\boldsymbol{\mu}) = \left|\det\left(G^{\text{aff},k}(\boldsymbol{\mu})\right)\right|, \quad 1 \leq k \leq K_{\text{dom}}, \tag{123}$$

which are constants in space over each subdomain. We further define, for any $\boldsymbol{\mu} \in \mathcal{D}$

$$D^{\text{aff},k}(\boldsymbol{\mu}) = G^{\text{aff},k}(\boldsymbol{\mu}), \tag{124}$$

this matrix shall prove convenient in subsequent *derivative* transformations, as we will see in section 4.3.
We may interpret our *local mappings* in terms of a *global transformation.* In particular, for any $\boldsymbol{\mu} \in \mathcal{D}$, the local mapping (120) induces a global *bijective piecewise-affine* mapping $\mathcal{T}^{\text{aff}} : \Omega_o(\boldsymbol{\mu}) \to \Omega_r$, such that:

$$\mathcal{T}^{\text{aff}}(\mathbf{x}_o; \boldsymbol{\mu}) = \mathcal{T}^{\text{aff},k}(\mathbf{x}_o; \boldsymbol{\mu}), \qquad k = \min_{k' \in \{1,\dots,K_{\text{dom}}\} \,|\, \mathbf{x} \in \Omega_o^{k'}(\boldsymbol{\mu})} \tag{125}$$

note the one-to-one property of this mapping (and, hence the arbitrariness of our *min* choice in (125)) is ensured by the interface condition (121).
In the following section the creation of an affine mappings will be discussed, subsequently the treatment of the parametric geometry dependence will be exploited thanks to an operative example.

4.2 Affine mappings construction

For simplicity we now consider a single subdomain, nonetheless the extension to the multi-subdomains case is readily obtainable as we will see. As we consider a single subdomain in this section, we shall suppress the subdomain superscript for clarity of exposition. The procedure, in the 2D case, is explained in Rozza et al. (2008), in this work it has been extended to the more general 3D case.
In the 3D case ($d = 3$, see equation 122) the affine mapping coefficients are $C^{\text{aff}}(\boldsymbol{\mu}) \in \mathbb{R}^3$ and $G^{\text{aff}}(\boldsymbol{\mu}) \in \mathbb{R}^{3\times 3}$, that is we have $3 + 9 = 12$ unknows to find in order to entirely define the affine transformation.
Under our assumption that the mapping is invertible, we know that the Jacobian J^{aff} of (123) is *strictly positive* and that the *derivative transformation matrix* D^{aff} of (124) is well defined.

Then the mapping coefficient can be identified by the relationship between 4 *non-planar parametrized image points* $\in \Omega_o(\boldsymbol{\mu})$, denoted with $\underline{\mathbf{z}}_o(\boldsymbol{\mu})$ and the corresponding 4 pre-image points $\in \Omega_r$, denoted with $\underline{\mathbf{z}}_r$ [6]:

$$\left\{\begin{array}{c} \underline{\mathbf{z}}_o^1(\boldsymbol{\mu}) \\ \underline{\mathbf{z}}_o^2(\boldsymbol{\mu}) \\ \underline{\mathbf{z}}_o^3(\boldsymbol{\mu}) \\ \underline{\mathbf{z}}_o^4(\boldsymbol{\mu}) \end{array}\right\} = \left\{\begin{array}{c} \{z_{o_1}^1, z_{o_2}^1, z_{o_3}^1\} \\ \{z_{o_1}^2, z_{o_2}^2, z_{o_3}^2\} \\ \{z_{o_1}^3, z_{o_2}^3, z_{o_3}^3\} \\ \{z_{o_1}^4, z_{o_2}^4, z_{o_3}^4\} \end{array}\right\},$$

$$\left\{\begin{array}{c} \underline{\mathbf{z}}_r^1 \\ \underline{\mathbf{z}}_r^2 \\ \underline{\mathbf{z}}_r^3 \\ \underline{\mathbf{z}}_r^4 \end{array}\right\} = \left\{\begin{array}{c} \{z_{r_1}^1, z_{r_2}^1, z_{r_3}^1\} \\ \{z_{r_1}^2, z_{r_2}^2, z_{r_3}^2\} \\ \{z_{r_1}^3, z_{r_2}^3, z_{r_3}^3\} \\ \{z_{r_1}^4, z_{r_2}^4, z_{r_3}^4\} \end{array}\right\}.$$

In particular, for given $\boldsymbol{\mu} \in \mathcal{D}$, the application of (122) to the selected nodes yields to:

$$z_{r_i}^m = C_i^{\text{aff}} + \sum_{j=1}^{3} G_{ij}^{m} \, z_{o_j}^{m}, \quad 1 \le i \le 3, \quad 1 \le m \le 4, \tag{126}$$

The (126) provides a system made of 12 equations, by which to determine the 12 *mapping coefficients.* If we choose at least two coplanar points, than the system is singular.
To be more explicit, we provide a matricial representation of equation 126:

$$\underline{\underline{\mathbb{B}}}^{\text{aff}} \underline{\mathbf{c}}^{\text{aff}} = \underline{\mathbf{v}}^{\text{aff}}. \tag{127}$$

[6] Here we denote with the superscript one of the 4 point considered, whereas the subscript indicates one of the 3 the components (x, y, z) of the spatial coordinates.

Where the matrix $\mathbb{B} \in \mathbb{R}^{12\times 12}$ summarizes the coefficients of the linear system, that depends upon the coordinates of the *"original"* points:

$$\underline{\underline{\mathbb{B}}}^{\text{aff}} = \begin{bmatrix} \begin{bmatrix} \\ \underline{\underline{I}}^{3\times 3} \\ \\ \end{bmatrix} & \begin{bmatrix} \underline{\mathbf{z}}_o^1 & \underline{0} & \underline{0} \\ \underline{0} & \underline{\mathbf{z}}_o^1 & \underline{0} \\ \underline{0} & \underline{0} & \underline{\mathbf{z}}_o^1 \end{bmatrix} \\ \begin{bmatrix} \\ \underline{\underline{I}}^{3\times 3} \\ \\ \end{bmatrix} & \begin{bmatrix} \underline{\mathbf{z}}_o^2 & \underline{0} & \underline{0} \\ \underline{0} & \underline{\mathbf{z}}_o^2 & \underline{0} \\ \underline{0} & \underline{0} & \underline{\mathbf{z}}_o^2 \end{bmatrix} \\ \begin{bmatrix} \\ \underline{\underline{I}}^{3\times 3} \\ \\ \end{bmatrix} & \begin{bmatrix} \underline{\mathbf{z}}_o^1 & \underline{0} & \underline{0} \\ \underline{0} & \underline{\mathbf{z}}_o^2 & \underline{0} \\ \underline{0} & \underline{0} & \underline{\mathbf{z}}_o^3 \end{bmatrix} \end{bmatrix} \tag{128}$$

moreover $\underline{\mathbf{c}}^{\text{aff}} \in \mathbb{R}^{12\times 1}$ is the array of unknows (i.e. the *mapping coefficients*) sorted as shown in (129); finally $\underline{\mathbf{v}}^{\text{aff}} \in \mathbb{R}^{12\times 1}$ is the array of known terms, that depends upon the coordinates of the *reference* points:

$$\underline{\mathbf{c}}^{\text{aff}} = \begin{Bmatrix} C_1^{\text{aff}} \\ C_2^{\text{aff}} \\ C_3^{\text{aff}} \\ G_{11}^{\text{aff}} \\ G_{12}^{\text{aff}} \\ G_{13}^{\text{aff}} \\ G_{21}^{\text{aff}} \\ G_{22}^{\text{aff}} \\ G_{23}^{\text{aff}} \\ G_{31}^{\text{aff}} \\ G_{32}^{\text{aff}} \\ G_{33}^{\text{aff}} \end{Bmatrix}, \quad \underline{\mathbf{v}}^{\text{aff}} = \begin{Bmatrix} \underline{\mathbf{z}}_r^1 \\ \underline{\mathbf{z}}_r^2 \\ \underline{\mathbf{z}}_r^3 \\ \underline{\mathbf{z}}_r^4 \end{Bmatrix}. \tag{129}$$

The mapping coefficients can be easily found solving the linear system (127) as it follows:

$$\underline{\mathbf{c}}^{\text{aff}} = \underline{\underline{\mathbb{B}}}^{\text{aff}^{-1}} \underline{\mathbf{v}}^{\text{aff}}. \tag{130}$$

The solution of the system requires 12^3 operation, negligible if compared to the previously discussed basis assembling cost.

Single domain mapping. We now use an example to illustrare the procedure. We will use as test case the transformation depicted in Figure 7.

With regard to the figure 7, we choose as geometrical parameters $\boldsymbol{\mu} = \{\mu_2, \mu_3, \mu_1\} = \{2, 3, 4\}$, in addition to simplify we choose $\underline{d}_r = \underline{d}_o(\boldsymbol{\mu})$, hence

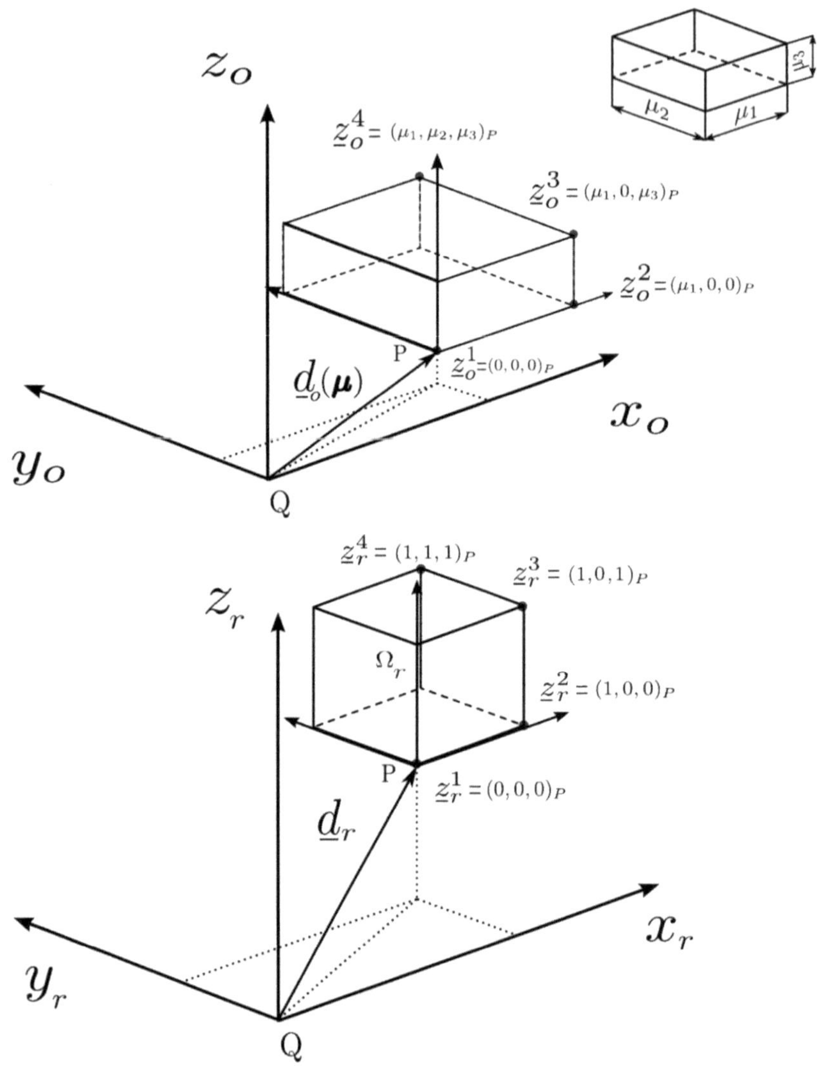

Figure 7: Affine transformation construction

we can use a local system attached to the first node ($\underline{\mathbf{z}}_*^1$).
Now, exploiting the procedure showed in section 4.2, we can build the system (127) by which we obtain the mapping coefficients.

$$
\begin{Bmatrix} C_1^{\text{aff}} \\ C_2^{\text{aff}} \\ C_3^{\text{aff}} \\ G_{11}^{\text{aff}} \\ G_{12}^{\text{aff}} \\ G_{13}^{\text{aff}} \\ G_{21}^{\text{aff}} \\ G_{22}^{\text{aff}} \\ G_{23}^{\text{aff}} \\ G_{31}^{\text{aff}} \\ G_{32}^{\text{aff}} \\ G_{33}^{\text{aff}} \end{Bmatrix} = \left(\begin{bmatrix}
1 & 0 & 0 & 0 & 0 & 0 & 0 & 0 & 0 & 0 & 0 & 0 \\
0 & 1 & 0 & 0 & 0 & 0 & 0 & 0 & 0 & 0 & 0 & 0 \\
0 & 0 & 1 & 0 & 0 & 0 & 0 & 0 & 0 & 0 & 0 & 0 \\
1 & 0 & 0 & \mu_1 & 0 & 0 & 0 & 0 & 0 & 0 & 0 & 0 \\
0 & 1 & 0 & 0 & 0 & 0 & \mu_1 & 0 & 0 & 0 & 0 & 0 \\
0 & 0 & 1 & 0 & 0 & 0 & 0 & 0 & 0 & \mu_1 & 0 & 0 \\
1 & 0 & 0 & \mu_1 & 0 & \mu_3 & 0 & 0 & 0 & 0 & 0 & 0 \\
0 & 1 & 0 & 0 & 0 & 0 & \mu_1 & 0 & \mu_3 & 0 & 0 & 0 \\
0 & 0 & 1 & 0 & 0 & 0 & 0 & 0 & 0 & \mu_1 & 0 & \mu_3 \\
1 & 0 & 0 & \mu_1 & \mu_2 & \mu_3 & 0 & 0 & 0 & 0 & 0 & 0 \\
0 & 1 & 0 & 0 & 0 & 0 & \mu_1 & \mu_2 & \mu_3 & 0 & 0 & 0 \\
0 & 0 & 1 & 0 & 0 & 0 & 0 & 0 & 0 & \mu_1 & \mu_2 & \mu_3
\end{bmatrix} \right)^{-1} \begin{Bmatrix} 0 \\ 0 \\ 0 \\ 1 \\ 0 \\ 0 \\ 1 \\ 0 \\ 1 \\ 1 \\ 1 \\ 1 \end{Bmatrix}. \tag{131}
$$

Then solving the system (131) we obtain:

$$
\underline{C}^{\text{aff}}(\boldsymbol{\mu}) = \begin{Bmatrix} 0 \\ 0 \\ 0 \end{Bmatrix}, \quad \underline{\underline{G}}^{\text{aff}}(\boldsymbol{\mu}) = \begin{bmatrix} \frac{1}{\mu_1} & 0 & 0 \\ 0 & \frac{1}{\mu_2} & 0 \\ 0 & 0 & \frac{1}{\mu_3} \end{bmatrix}. \tag{132}
$$

The Jacobian of the transformation (123) is $J^{\text{aff}}(\boldsymbol{\mu}) = \frac{1}{\mu_1\mu_2\mu_3} = \frac{1}{12}$.
We remark that the Jacobian of a transformation can be seen as the ratio between the final and initial volumes on which the deformation takes place. In order to verify the affine transformation we apply the transformation to each nodes of the *original domain* and, the *affine mapping* should trace back the corresponding nodes on the *reference domain*:

$$
\begin{aligned}
&\boxed{\mathbf{z}^1}\ \underline{\underline{G}}^{\text{aff}} \begin{Bmatrix} 0 \\ 0 \\ 0 \end{Bmatrix} = \begin{Bmatrix} 0 \\ 0 \\ 0 \end{Bmatrix}, \qquad \boxed{\mathbf{z}^2}\ \underline{\underline{G}}^{\text{aff}} \begin{Bmatrix} \mu_1 \\ 0 \\ 0 \end{Bmatrix} = \begin{Bmatrix} 1 \\ 0 \\ 0 \end{Bmatrix}, \\
&\boxed{\mathbf{z}^3}\ \underline{\underline{G}}^{\text{aff}} \begin{Bmatrix} \mu_1 \\ 0 \\ \mu_3 \end{Bmatrix} = \begin{Bmatrix} 1 \\ 0 \\ 1 \end{Bmatrix}, \qquad \boxed{\mathbf{z}^4}\ \underline{\underline{G}}^{\text{aff}} \begin{Bmatrix} \mu_1 \\ \mu_2 \\ \mu_3 \end{Bmatrix} = \begin{Bmatrix} 1 \\ 1 \\ 1 \end{Bmatrix}.
\end{aligned} \tag{133}
$$

We remark that in this simple test case the transformation is *"diagonal"* because we are deforming the domain by mean of a simple *dilation*. In the case of a shear deformation for example, we would have even the *extradiagonal* terms (Rozza et al., 2008).

Global affine mappings. Exploited the case of a single domain, we need to extend the procedure to a multi-subdomain case. We will make use of an example sketched in Figure 8. We have two adjacent subdomains denoted

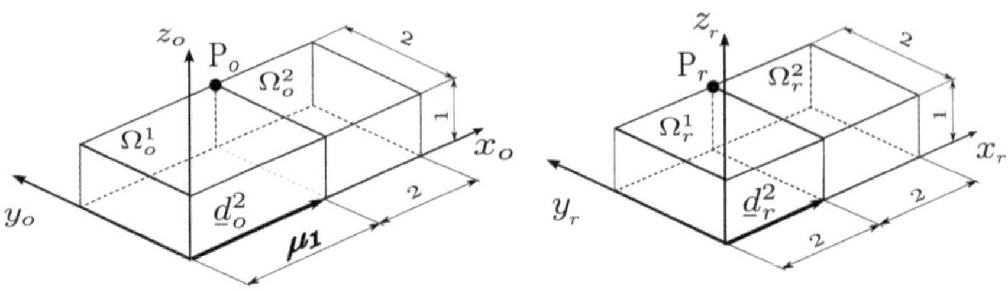

Figure 8: Global affine mappings

with Ω_*^1 and Ω_*^2. The first domain can be deformed along the x-axes thanks to the parameter denoted with μ_1, all the other dimensions are held fixed. The procedure is quite similar, we just need in addition to satisfy the *global continuity condition* (121). A way to satisfy it, is to use a *unique system of reference* for the different subdomains.
In this way the procedure described in section 4.2 will *implicitly* provide the suitable translation $\underline{C}^{\text{aff}}(\boldsymbol{\mu})$ to satisfy the interface condition.
Using the procedure for the two domains we obtain the following results:

$$\begin{aligned}
\underline{C}_1^{\text{aff}}(\boldsymbol{\mu}) &= \left\{ \begin{array}{c} 0 \\ 0 \\ 0 \end{array} \right\}, \qquad \underline{\underline{G}}_1^{\text{aff}}(\boldsymbol{\mu}) = \left[\begin{array}{ccc} \frac{1}{\mu_1} & 0 & 0 \\ 0 & 1 & 0 \\ 0 & 0 & 1 \end{array} \right], \\
\underline{C}_2^{\text{aff}}(\boldsymbol{\mu}) &= \left\{ \begin{array}{c} 2-\mu_1 \\ 0 \\ 0 \end{array} \right\}, \qquad \underline{\underline{G}}_2^{\text{aff}}(\boldsymbol{\mu}) = \left[\begin{array}{ccc} 1 & 0 & 0 \\ 0 & 1 & 0 \\ 0 & 0 & 1 \end{array} \right].
\end{aligned} \tag{134}$$

We now take two *adjacent nodes* (denoted with P_o in the Figure 8) on the *original domains*; if the global mappings satisfies the *interface condition*, applying the two different *affine transformation* at the node, we will obtain again two *adjacent nodes* (P_r) on the *reference domain*:

$$\mathcal{T}_1^{\text{aff}}(P_o(\boldsymbol{\mu}),\boldsymbol{\mu}) = \underline{C}_1^{\text{aff}} + \underline{\underline{G}}_1^{\text{aff}}\ \{\mu_1, 2, 1\}^T = \{2, 2, 1\} = P_r,$$

$$\mathcal{T}_2^{\text{aff}}(P_o(\boldsymbol{\mu}),\boldsymbol{\mu}) = \underline{C}_2^{\text{aff}} + \underline{\underline{G}}_2^{\text{aff}}\ \{\mu_1, 2, 1\}^T = \{2, 2, 1\} = P_r.$$

The node P_o is identically projected into P_r thanks to the two different *affine transformations*, hence the *continuity* of the *global mapping* is satisfied.

4.3 Linear and bilinear forms

We now focus on the transformations that we have to operate on the *weak forms* that arise from our system of partial differential equations, if our domain $\Omega_o(\boldsymbol{\mu})$ allows the *affine geometry precondition* described in the previous section.
We will use a simple *scalar coercive* 3D problem to show how to exploit the *geometry parametric dependence* of the domain.
The procedure is discussed in the 2D case in Rozza et al. (2008); Rozza (2009a).

Formulation on the original domain. The problem is initially posed on the original domain $\Omega_o(\boldsymbol{\mu})$. We shall assume for simplicity that $X_0^e(\boldsymbol{\mu}) = H_0^1(\Omega_0(\boldsymbol{\mu}))$, which corresponds to homogeneous Dirichlet boundary conditions over the entire boundary $\partial\Omega_o(\boldsymbol{\mu})$. Given $\boldsymbol{\mu} \in \mathcal{D}$, find $u_o^e(\boldsymbol{\mu}) \in X_0^e(\boldsymbol{\mu})$ such that:

$$a_o(u_o^e(\boldsymbol{\mu}), v; \boldsymbol{\mu}) = f_o(v; \boldsymbol{\mu}), \quad \forall v \in X_0^e(\boldsymbol{\mu}) \tag{135}$$

then evaluate

$$s_o^e(\boldsymbol{\mu}) = f_o(u_o^e(\boldsymbol{\mu})). \tag{136}$$

We now place conditions on a_o and f_o such that, in conjunction with the *affine geometry precondition*, we are ensured an *affine expansion* of the bilinear form (11).
We require that $a_o(\cdot\,, \cdot\,; \boldsymbol{\mu}) : H_0^1(\Omega_o(\boldsymbol{\mu})) \times H_0^1(\Omega_o(\boldsymbol{\mu})) \to \mathbb{R}$ can be expressed as:

$$a_o(w, v; \boldsymbol{\mu}) =$$

$$= \sum_{k=1}^{K_{\mathrm{dom}}} \int_{\Omega_o^k(\boldsymbol{\mu})} \left\{ \frac{\partial w}{\partial x_{o_1}} \quad \frac{\partial w}{\partial x_{o_2}} \quad \frac{\partial w}{\partial x_{o_3}} \quad w \right\} \mathcal{K}_{o_{ij}}^k(\boldsymbol{\mu}) \left\{ \begin{array}{c} \dfrac{\partial w}{\partial x_{o_1}} \\ \dfrac{\partial w}{\partial x_{o_2}} \\ \dfrac{\partial w}{\partial x_{o_3}} \\ w \end{array} \right\} d\Omega_o^k \tag{137}$$

where $\mathbf{x}_o = \{x_{o_1}, x_{o_2}, x_{o_3}\}$ denotes a point in $\Omega_o(\boldsymbol{\mu})$; and where for $1 \leq k \leq K_{\mathrm{dom}}$, $\underline{\underline{\mathcal{K}}}_o^k : \mathcal{D} \to \mathbb{R}^{4\times 4}$ is a given *symmetric positive definite* matrix (which in turn *ensures coercivity* of our bilinear form):

$$\underline{\underline{\mathcal{K}}}_o^k = \begin{bmatrix} \begin{bmatrix} \mathbb{R}^{3\times 3} \\ \\ \mathbb{R}^{1\times 3} \end{bmatrix} & \begin{bmatrix} \mathbb{R}^{3\times 1} \\ \\ \mathbb{R}^{1\times 1} \end{bmatrix} \end{bmatrix} \tag{138}$$

the upper 3×3 principal submatrix of $\underline{\underline{\mathcal{K}}}_o^k$ represent the usual *diffusion/conductivity tensor*, the element $(4,4)$ represent the identity tensor (*mass matrix* or a *reaction term*), finally the elements $\underline{\underline{\mathcal{K}}}^k_{o_{1:3,3}}$ and $\underline{\underline{\mathcal{K}}}^k_{o_{3,1:3}}$, that we set to zero because we are dealing with symmetric operators, represent the *first derivative* operators (i.e. *convective terms*).
Similarly we require that $f_o : H_o^1(\Omega_o(\boldsymbol{\mu})) \to \mathbb{R}$ can be expressed as

$$f_o(v;\boldsymbol{\mu}) = \sum_{k=1}^{K_{\text{dom}}} \int_{\Omega_o^k(\boldsymbol{\mu})} \underline{\mathcal{F}}_o^k \, v \, d\Omega_o^k. \tag{139}$$

In this case we have assumed that the linear functional is only due to *volume force*, a similar treatment is possible in the case of *Dirichlet non-homogeneus* condition and/or *non-homogeneus Neumann* condition.

Formulation on reference domain. We now apply standard techniques to transform the problem over the original domain to an equivalent problem over the reference domain.
Given $\boldsymbol{\mu} \in \mathcal{D}$, find $u^e(\boldsymbol{\mu}) \in X^e \equiv H_0^1(\Omega)$ such that:

$$a(u^e, v; \boldsymbol{\mu}) = f(v;\boldsymbol{\mu}) \quad \forall v \in X^e \tag{140}$$

then evaluate

$$s^e(\boldsymbol{\mu}) = f(u^e(\boldsymbol{\mu})). \tag{141}$$

We may then identify the relations between the *output* and the *solution field*, in the original and in the reference domain:

$$\begin{aligned} s^e(\boldsymbol{\mu}) &= s_o^e(\boldsymbol{\mu}) \\ u^e(\boldsymbol{\mu}) &= u_o^e(\boldsymbol{\mu}) \circ \mathcal{T}^{\text{aff}}(\,\cdot\,;\boldsymbol{\mu}). \end{aligned} \tag{142}$$

The transformed bilinear form a, can be expressed as:

$$a(w,v;\boldsymbol{\mu}) = \tag{143}$$

$$\int_{\Omega_r^k} \left\{ \frac{\partial w}{\partial x_{r_1}} \quad \frac{\partial w}{\partial x_{r_2}} \quad \frac{\partial w}{\partial x_{r_3}} \quad w \right\} \mathcal{K}_{ij}^k(\boldsymbol{\mu}) \left\{ \begin{array}{c} \dfrac{\partial w}{\partial x_{r_1}} \\ \dfrac{\partial w}{\partial x_{r_2}} \\ \dfrac{\partial w}{\partial x_{r_3}} \\ w \end{array} \right\} d\Omega_r^k \tag{144}$$

where $\mathbf{x}_r = \{x_{r_1}, x_{r_2}, x_{r_3}\}$ denotes a point in Ω_r and where $\mathcal{K}\|_r : \mathcal{D} \to \mathbb{R}^{4\times4}$, $1 \le k \le K_{\text{dom}}$ are symmetric positive definite matrices.
To obtain this matrices we first need to find the relation between the *derivative operator* written in the original domain and the corresponding operator written in the original domain. In particular, we have that

$$\frac{\partial\,\cdot}{\partial x_{o_i}} = \frac{\partial x_{r_j}}{\partial x_{o_i}} \frac{\partial\,\cdot}{\partial x_{r_j}} = G_{ij}^{\text{aff}}(\boldsymbol{\mu}) \frac{\partial\,\cdot}{\partial x_{o_i}} = \underline{\underline{D}}^{\text{aff}}(\boldsymbol{\mu}) \frac{\partial\,\cdot}{\partial x_{o_i}}. \tag{145}$$

The definition (124) of the *derivatives operator* $\underline{\underline{D}}^{\text{aff},k}$ is now clear; the matrices $\underline{\underline{G}}^{\text{aff},k}$ automatically provide the relation between the derivatives operator in the original and in the reference domain.
Moreover, since we are acting a change of variable $\mathbf{x}_o \to \mathbf{x}_r$ in the integral (137), recalling the equation (123), we get

$$d\Omega_o(\boldsymbol{\mu}) = \det(\underline{\underline{G}}^{\text{aff}}(\boldsymbol{\mu})^{-1}) d\Omega_r = \left(J^{\text{aff},k}(\boldsymbol{\mu})\right)^{-1}. \tag{146}$$

It follows that, considering the equations (145)-(146), the relation between $\underline{\underline{\mathcal{K}}}_o^k(\boldsymbol{\mu})$ and $\underline{\underline{\mathcal{K}}}_r(\boldsymbol{\mu})$ can be written as

$$\underline{\underline{\mathcal{K}}}_r^k(\boldsymbol{\mu}) = \left(\underline{\underline{\mathcal{G}}}^k(\boldsymbol{\mu})\right)^T \underline{\underline{\mathcal{K}}}_o^k(\boldsymbol{\mu})\, \underline{\underline{\mathcal{G}}}^k(\boldsymbol{\mu})\, \left(J^{\text{aff},k}(\boldsymbol{\mu})\right)^{-1} \quad 1 \le k \le K_{\text{dom}} \tag{147}$$

where we have defined $\underline{\underline{\mathcal{G}}}^k(\boldsymbol{\mu}) : \mathcal{D} \to \mathbb{R}^{4\times4}$, $1 \le k \le K_{\text{dom}}$ as

$$\underline{\underline{\mathcal{G}}}^k(\boldsymbol{\mu}) = \begin{bmatrix} \underline{\underline{\mathcal{D}}}^{\text{aff},k}(\boldsymbol{\mu}) & \underline{0}^{3\times1} \\ \underline{0}^{1\times3} & 1 \end{bmatrix}. \tag{148}$$

Similarly, the transformed linear form can be expressed as

$$f(v;\boldsymbol{\mu}) = \sum_{k=1}^{K_{\text{dom}}} \int_{\Omega_r^k} \mathcal{F}^k(\boldsymbol{\mu})\, v\, d\Omega_r^k. \tag{149}$$

Here $\mathcal{F}^k(\boldsymbol{\mu}) : \mathcal{D} \to \mathbb{R}$, $1 \le k \le K_{\text{dom}}$ is given by:

$$\mathcal{F}^k(\boldsymbol{\mu}) = \mathcal{F}_o^k \left(J^{\text{aff},k}(\boldsymbol{\mu})\right)^{-1} \quad 1 \le k \le K_{\text{dom}}. \tag{150}$$

We note that, in general, the $\underline{\underline{\mathcal{K}}}^k(\boldsymbol{\mu})$, $\mathcal{F}(\boldsymbol{\mu})$, will be different for each subdomain Ω^k, $1 \le k \le K_{\text{dom}}$. The differences can be due to *property variation* (e.g. a diffusivity of a particular subdomain) or to *geometry variation* (e.g. a characteristic dimension of the physical problem), or *both.*
We thus require, as already indicated earlier, that the *FE* approximation be built upon a *subtriangulation* of the *RB* discretization: discontinuities in *PDEs* coefficients are therefore restricted to element faces.
In this way, the boundary elements chosen for the *RB* triangulation will delimit a very well defined region of space (our *RB* subdomains), on which we assume that the parameters will be *constants in space.* This allows a simpler *identification/extraction* of the terms in the affine expansion (11), as we now discuss.

Affine form. We focus here on a, though f admits a similar treatment. We simply expand the transformed form on the reference domain, by considering in turn each subdomain Ω_r^k and *each entry* of the diffusivity/conductivity tensor $\mathcal{K}_{ij}$, $1 \le ij \le 4$, $1 \le k \le K_{\text{dom}}$.
Thus the affine form (144) can be written as follows:

$$\begin{aligned} a(w,v;\boldsymbol{\mu}) =& \mathcal{K}_{11}^1 \int_{\Omega_r^1} \frac{\partial w}{\partial x_{r_1}} \frac{\partial v}{\partial x_{r_1}} \, d\Omega_r^1 + \dots \\ & \dots + \underbrace{\mathcal{K}_{ij}^k}_{\theta_a^q(\boldsymbol{\mu})} \underbrace{\int_{\Omega_r^k} \frac{\partial w}{\partial x_{r_i}} \frac{\partial v}{\partial x_{r_j}} \, d\Omega_r^k}_{a^q(w,v)} + \dots \\ & \dots + \mathcal{K}_{44}^{K_{\text{dom}}} \int_{\Omega_r^{K_{\text{dom}}}} w\, v \, d\Omega_r^{K_{\text{dom}}}. \end{aligned} \tag{151}$$

We can then identify each component in the affine expansion: for each term in 151 the pre-factor of the integral represents $\theta_a^q(\boldsymbol{\mu})$, whereas the integral represents the parameter independent matrices $\underline{\underline{\mathbb{A}}}^q = a^q(w,v)$.
For a better understanding of what we have just obtained, we can have a look to the equation (69), in which we were building the *RB* system in the coercive case. The *parameter independent* matrices $\underline{\underline{\mathbb{A}}}^q$ can be now exploited even in the general case of *geometric parameter dependence*, thanks to the *geometric affine precondition.*

Affine expansion terms count. In the most general *scalar* case, the number of affine expansion terms can be (at most) $Q_a = 4 \times 4 \times K_{\text{dom}}$. Exploiting the symmetry of the bilinear form, hence of the tensor $\underline{\underline{\mathcal{K}}}^k$, only $Q_a = 10 \times K_{\text{dom}}$ terms are needed. In fact since $\mathcal{K}^k_{ij} = \mathcal{K}^k_{ji}$, $i \neq j$, the pre-factor associated to these integrals can be assembled together.
We first consider the 6 differents entries of the simmetric tensor $\mathcal{K}$ of the first subdomain, then the second subdomain and so on. Hence, the $\theta^q_a(\boldsymbol{\mu})$ and the associated *parametric independent matrices* are given by Table 3.

q	$\theta^q_f(\boldsymbol{\mu})$	$\mathbb{F}^q$
1	$\mathcal{K}_{11}$	$\displaystyle\int_{\Omega^1_r} \frac{\partial w}{\partial x_{r_1}} \frac{\partial v}{\partial x_{r_1}} \, d\Omega^1_r$
2	$\mathcal{K}^1_{12}$	$\displaystyle\int_{\Omega^1_r} \frac{\partial w}{\partial x_{r_1}} \frac{\partial v}{\partial x_{r_2}} \, d\Omega^1_r$
7	$\mathcal{K}^2_{11}$	$\displaystyle\int_{\Omega^2_r} \frac{\partial w}{\partial x_{r_1}} \frac{\partial v}{\partial x_{r_1}} \, d\Omega^2_r$
Q_a	$\mathcal{K}^{K_{\text{dom}}}_{11}$	$\int_{\Omega^{K_{\text{dom}}}_r} w\, v \, d\Omega^{K_{\text{dom}}}_r$

Table 3: $\theta^q(\boldsymbol{\mu})$–functions and parameter independent matrices

Dealing with the *vectorial* case, the number of affine expansion terms can be $Q_a = 9 \times 9 \times K_{\text{dom}}$. Therefore it is crucial, in order to reduce the *RB* computational cost that depends on Q_a (see Table 3), to minimize the number of terms of the affine expansion.

5 Thermal problem

In this section we will exploit the creation of our *RB* example *in the 3D case* dealing with a *scalar elliptic coercive* problem.
We deal with a steady conduction thermal problem, assuming that the thermal conductivity $\underline{\underline{K}}$ is represented by a positive definite matrix; then the unknown is the field of temperature, that we will denote with $u(\mathbf{x}; \boldsymbol{\mu}) \in \mathbb{R}$. This class of problem, although rather simple, is able to describe a wide range of continuum mechanics problems, see for example Arpaci (1966).
We mention, for instance, the study of the performance of an heat sink designed for the thermal management of high-density electronic components, the design of an insulated coverage of a building to reduce the energetic

consumption, the control of the temperature within an engine shaft to prevent thermal stresses or deformation, etc...
Another notable application can be the non-destructive testing of mechanical components or the identification of an inclusion within a casting steel; in short, despite the simple mathematical fomrulation, this case is more than a mathematical abstraction.

5.1 Problem description

We now briefly introduce the Thermal Block problem, henceforth we will refer to this case with the "TB" label.

Physical problem. The problem of a *steady-conduction* is considered here in a cubic domain.
We want to evaluate the thermal field in a non-isotropic conductive block, with an heat flux q imposed on a face of the cube, with regard to a reference environmental temperature taken on the opposite face and assuming that the other faces are insulated. The cube has an anisotropic conductivity due to an inclusion of different materials within the piece.
Dealing for example with the material science, the inclusion can be (Walker, 1993):

1. a *deficiency* of material, a hole due to gas bubble present within the casting during solidification, or due to a fatigue crack;
2. an inhomogeneity of the material, due to an unbalancing of concentration of the alloyants.

Therefore, in order to seize the behavior of such a phenomena, the *RB* parametric approach is very well suited. The thermal problem, in the 2D frame, has been treated in Rozza et al. (2008); Patera and Rozza (2007) and Huynh et al. (2007).
The output of interest is the average temperature of the heated face.
Now we formulate the problem into the mathematical setting.

Analytical problem. The equation that describes the field of temperature is represented by the *Fourier equation* in which the time derivative will be neglected thanks to the hypothesis of a steady state problem (no transient effect considered) as well as the volume force that will be considered negligible.

5.2 Parameters

Due to the nature of the inclusion, its shape and dimension can vary arbitrarily, hence we introduce some geometric parameters in order to de-

scribe it. It is important to remark that, due the fact that the parametrized geometry should allow an *affine geometric precondition* (discussed in section 4), we restrict our attention to a parallelepiped inclusion.
In particular, with regard to the Figure 9, we have assumed that the cubic block can vary its *dimension* along the three axis by means of a set of parameters $\boldsymbol{\mu} = \{\mu_4, \mu_5, \mu_6\}$. In addition we have chosen other three parameters, denoted by $\boldsymbol{\mu} = \{\mu_1, \mu_2, \mu_3\}$ to parametrize the *position* of the inclusion. The solid is sketched as $3 \times 3 \times 3$ cubic blocks, the central block

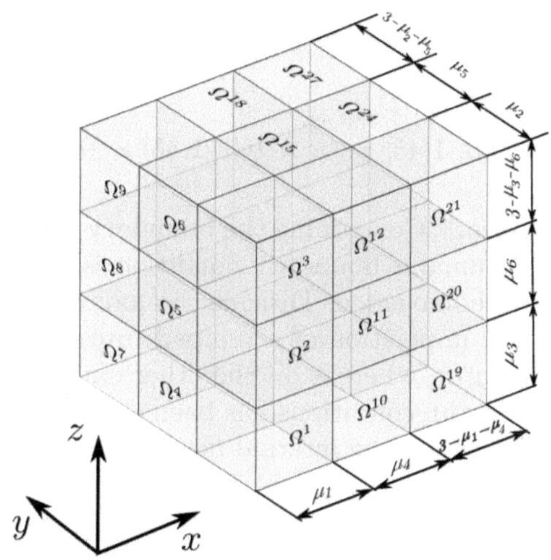

Figure 9: TB domain decomposition

(subdomain Ω^{14} in our scheme) represents the inclusion.
Each block, due to the parametric geometry dependence of the inclusion and to the hypothesis required by an affine geometry (Section 4.1), is subjected to a geometrical parametric dependence. Each sub-block is considered *isotropic.* The conductivity constant for the central sub-block (inclusion) is another parameter, denoted by μ_7, whereas for the other sub-blocks the conductivity is the unity (reference). We now summarize the parameters and the parameter domain chosen to describe our *TB* problem as:

- the dimension and the position of the inclusion, that can be described by 6 parameters, that are 3 *translations* and 3 *dimensions.* The parameters $\mu_{1:6}$ are shown in Figure 9.

The parameter domain for the geometrical quantities is:

$$\mathcal{D}_{\text{geom}} = \left[\mu_1^{\min}, \mu_1^{\max}\right] \times \ldots \times \left[\mu_6^{\min}, \mu_6^{\max}\right]$$
$$= [0.5,\ 1.45] \times \ldots \times [0.5,\ 1.45]\,, \qquad (152)$$

- the conductivity coefficient of the inclusion, denoted with μ_7. The parameter domain for this physical quantity is:

$$\mathcal{D}_{\text{physics}} = \left[\mu_7^{\min}, \mu_7^{\max}\right] = [0.1,\ 10]\,. \qquad (153)$$

The parameter domain is therefore given by $\boldsymbol{\mu} \in \mathcal{D} \in \mathbb{R}^{P=7}$, such that:

$$\mathcal{D} = \mathcal{D}_{\text{geom}} \times \mathcal{D}_{\text{physics}}$$
$$= [0.5,\ 1.45] \times \ldots \times [0.5,\ 1.45] \times [0.1,\ 10]\,.$$

Boundary conditions. Concerning the boundary conditions (Figure 10), a non-homogeneus Neumann boundary condition is imposed on Γ_6 representing a *heat flux*, an homogeneus Dirichlet boundary condition is imposed on Γ_1 representing the imposition of a temperature (adimensional, i.e. environmental temperature), whereas on the other external faces of the cube $\Gamma_{2:5}$ homogeneus Neumann conditions has been chosen, representing insulation of the walls. Finally, on the internal faces we have assumed continuity of *temperature* and *fluxes*.

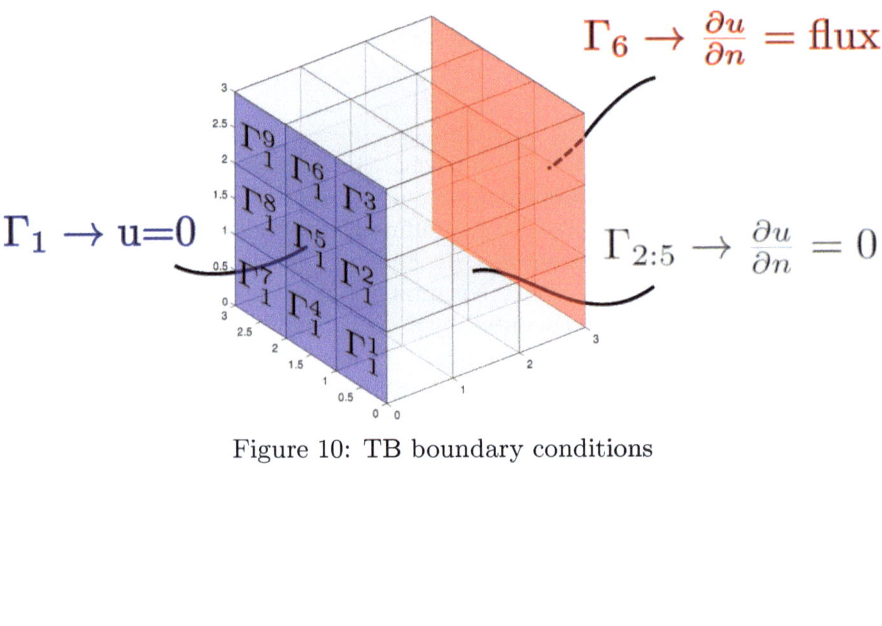

Figure 10: TB boundary conditions

5.3 TB Problem formulation

Original domain. We introduce the analytical formulation of the governing *PDEs* on the *original domain*. The equation which describes the field of temperature, within the hypothesis described in section 5, is the following:

$$\begin{cases} -\nabla \cdot \left(\underline{\underline{K}}_o \nabla_o u \right) = 0 & \text{in } \Omega_o(\boldsymbol{\mu}) \\ u = 0 & \text{on } \Gamma_{o_1}(\boldsymbol{\mu}) \\ \dfrac{\partial u}{\partial \mathbf{n}} = q & \text{on } \Gamma_{o_6}(\boldsymbol{\mu}) \end{cases} \tag{154}$$

Multiplying the equation by a suitable *test function* v such that $v \in X^e \equiv \{v \in H_0^1(\Omega) | \, v|_{\Gamma^D} = 0\}$ and integrating over the domain Ω we obtain:

$$\int_{\Omega_o(\boldsymbol{\mu})} -\nabla_{{}_o} \cdot \left(\underline{\underline{K}}_o \nabla_o u \right) v \, d\Omega_o = 0. \tag{155}$$

The domain Ω_o is the *original domain* on which the *PDE* is defined. The whole domain $\Omega_o(\mu)$ is decomposed (section 4.1) in K_{dom} non-overlapping subdomains such that:

$$\Omega_o(\mu) = \bigcup_{k=1}^{K_{27}} \Omega_o^k(\mu),$$

in addition, recalling the *Green Theorem* for the laplacian, (Quarteroni, 2013):

$$\int_\Omega \Delta u \, v \, d\Omega = \int_\Omega \nabla u \cdot \nabla v \, d\Omega - \int_{\partial\Omega} \frac{\partial u}{\partial \mathbf{n}} v \, d\gamma$$

the equation (155) becomes

$$\sum_{k=1}^{27} \int_{\Omega_o^k} -\underline{\underline{K}}_o \nabla_o u \cdot \nabla_o v \, d\Omega_o^k + \sum_{k=1}^{27} \int_{\partial\Omega_o^k} \underline{\underline{K}}_o \frac{\partial u}{\partial \mathbf{n}} v \, \partial\Omega_o^k = 0.$$

Thanks to the functional space chosen, the boundary terms vanish on the face on which we have imposed a zero temperature (on $\Gamma_1 \rightarrow v = 0$). The boundary terms relatives to $\Gamma_{2:5}$ vanish because we have imposed an homogeneus Neumann condition (on $\Gamma_{2:5} \rightarrow \partial u / \partial \mathbf{n} = 0$).
The internal faces contributes disappear thanks to the continuity of temperature and fluxes. Then the only remaining boundary term is the one relative to the face Γ_6 on which we have imposed an heat flux q.
Therefore, the equation (156) can be simplified as:

$$\sum_{k=1}^{27} \int_{\Omega_o^k} \underline{\underline{K}}_o \nabla_o u \cdot \nabla_o v \, d\Omega_o^k = \int_{\Gamma_{o_6}} \underline{\underline{K}}_o \frac{\partial u}{\partial \mathbf{n}} v \, d\Gamma_{o_6}. \tag{156}$$

Replacing the Neumann boundary condition (154):

$$\underline{\underline{K}}_o \nabla_o u \cdot n = \underline{\underline{K}}_o \frac{\partial u}{\partial \mathbf{n}} = q$$

in the *weak formulation* (156), we finally obtain:

$$\sum_{k=1}^{27} \int_{\Omega_o^k} \underline{\underline{K}}_o \nabla_o u \cdot \nabla_o v \, d\Omega_o^k = \int_{\Gamma_{o6}} q \, v \, d\Gamma_{o6}. \tag{157}$$

Introducing the bilinear form

$$a(u, v; \boldsymbol{\mu}) = \sum_{k=1}^{27} \int_{\Omega_o^k} \underline{\underline{K}}_o \nabla_o u \cdot \nabla_o v \, d\Omega_o^k \tag{158}$$

and the linear functional

$$f(v; \boldsymbol{\mu}) = \int_{\Gamma_{o6}} q \, v \, d\Gamma_{o6}, \tag{159}$$

we can restate the problem (157) as: find $u \in X^e(\Omega_o(\boldsymbol{\mu}))$, such that

$$a(u, v; \boldsymbol{\mu}) = f(v; \boldsymbol{\mu}) \quad \forall v \in X^e(\Omega_o(\boldsymbol{\mu})). \tag{160}$$

The coercivity and the continuity of the bilinear form a and the continuity of the functional f can be proved. Then the *Lax-Milgram* theorem ensures the existence and uniqueness of the solution, see Quarteroni (2013).

Reference domain. In this section we apply standard techniques to transform the problem statement over the *original domain* to an equivalent problem fomrulated over *reference domain.*
We shall be ultimately able to write the problem in an *affine formulation* (11), to exploit the crucial Offline/Online computational splitting procedure. In order to obtain the problem formulation (160) on the reference domain, we need to evaluate the *affine transformation* for each subdomain $\Omega_o^k(\boldsymbol{\mu}) \in \Omega_o(\boldsymbol{\mu})$, tracing back the *derivatives operator* and all the *geometric parameter dependent quantities* to the *reference domain* by the recipe provided in section 4.3.
In order to build the *affine decomposition*, we must compute the affine mappings for each subdomain $\mathcal{T}^{\mathrm{aff},k}(\,\cdot\,; \boldsymbol{\mu}) : \Omega_o(\boldsymbol{\mu}) \to \Omega_r$, $1 \le k \le 27$, in order to evaluate:

1. the Jacobian $J^{\mathrm{aff},k}(\boldsymbol{\mu})$, $1 \le k \le 27$ (123)

$$J^{\mathrm{aff},k}(\boldsymbol{\mu}) = \left|\det\left(\underline{\underline{G}}^{\mathrm{aff},k}(\boldsymbol{\mu})\right)\right|;$$

2. the derivatives operator $D^{\text{aff},k}(\boldsymbol{\mu})$ (equation 124)

$$\underline{\underline{D}}^{\text{aff},k}(\boldsymbol{\mu}) = \underline{\underline{G}}^{\text{aff},k}(\boldsymbol{\mu});$$

We will not present in detail the procedure to obtain the affine transformation for all the subdomains, we refer the reader to the section 4.2 for a detailed abstract explanation. In this section we just provide the results in Table 4.

We remark that the matrices $\underline{\underline{G}}^{\text{aff},k}(\boldsymbol{\mu})$, $1 \le k \le 27$ are diagonal thanks to the particular choice of the geometric parameters. Once the affine mappings

sub	G_{11}^{aff}	G_{22}^{aff}	G_{33}^{aff}	C_1^{aff}	C_2^{aff}	C_3^{aff}
Ω^1	$1/\mu_1$	$1/\mu_2$	$1/\mu_3$	0	0	0
Ω^2	$1/\mu_1$	$1/\mu_2$	$1/\mu_6$	0	0	$1-\mu_3$
Ω^3	$1/\mu_1$	$1/\mu_2$	$1/(3-\mu_6-\mu_3)$	0	0	$2-\mu_6-\mu_3$
Ω^4	$1/\mu_1$	$1/\mu_4$	$1/\mu_3$	0	$1-\mu_1$	0
Ω^5	$1/\mu_1$	$1/\mu_4$	$1/\mu_6$	0	$1-\mu_1$	$1-\mu_3$
Ω^6	$1/\mu_1$	$1/\mu_4$	$1/(3-\mu_6-\mu_3)$	0	$1-\mu_1$	$2-\mu_6-\mu_3$
Ω^7	$1/\mu_1$	$1/(3-\mu_5-\mu_2)$	$1/\mu_3$	0	$2-\mu_5-\mu_1$	0
Ω^8	$1/\mu_1$	$1/(3-\mu_5-\mu_2)$	$1/\mu_6$	0	$2-\mu_5-\mu_1$	$1-\mu_3$
Ω^9	$1/\mu_1$	$1/(3-\mu_5-\mu_2)$	$1/(3-\mu_6-\mu_3)$	0	$2-\mu_5-\mu_1$	$2-\mu_6-\mu_3$
Ω^{10}	$1/\mu_5$	$1/\mu_2$	$1/\mu_3$	$1-\mu_2$	0	0
Ω^{11}	$1/\mu_5$	$1/\mu_2$	$1/\mu_6$	$1-\mu_2$	0	$1-\mu_3$
Ω^{12}	$1/\mu_5$	$1/\mu_2$	$1/(3-\mu_6-\mu_3)$	$1-\mu_2$	0	$2-\mu_6-\mu_3$
Ω^{13}	$1/\mu_5$	$1/\mu_4$	$1/\mu_3$	$1-\mu_2$	$1-\mu_1$	0
Ω^{14}	$1/\mu_5$	$1/\mu_4$	$1/\mu_6$	$1-\mu_2$	$1-\mu_1$	$1-\mu_3$
Ω^{15}	$1/\mu_5$	$1/\mu_4$	$1/(3-\mu_6-\mu_3)$	$1-\mu_2$	$1-\mu_1$	$2-\mu_6-\mu_3$
Ω^{16}	$1/\mu_5$	$1/(3-\mu_5-\mu_2)$	$1/\mu_3$	$1-\mu_2$	$2-\mu_5-\mu_1$	0
Ω^{17}	$1/\mu_5$	$1/(3-\mu_5-\mu_2)$	$1/\mu_6$	$1-\mu_2$	$2-\mu_5-\mu_1$	$1-\mu_3$
Ω^{18}	$1/\mu_5$	$1/(3-\mu_5-\mu_2)$	$1/(3-\mu_6-\mu_3)$	$1-\mu_2$	$2-\mu_5-\mu_1$	$2-\mu_6-\mu_3$
Ω^{19}	$1/(3-\mu_4-\mu_1)$	$1/\mu_2$	$1/\mu_3$	$2-\mu_4-\mu_2$	0	0
Ω^{20}	$1/(3-\mu_4-\mu_1)$	$1/\mu_2$	$1/\mu_6$	$2-\mu_4-\mu_2$	0	$1-\mu_3$
Ω^{21}	$1/(3-\mu_4-\mu_1)$	$1/\mu_2$	$1/(3-\mu_6-\mu_3)$	$2-\mu_4-\mu_2$	0	$2-\mu_6-\mu_3$
Ω^{22}	$1/(3-\mu_4-\mu_1)$	$1/\mu_4$	$1/\mu_3$	$2-\mu_4-\mu_2$	$1-\mu_1$	0
Ω^{23}	$1/(3-\mu_4-\mu_1)$	$1/\mu_4$	$1/\mu_6$	$2-\mu_4-\mu_2$	$1-\mu_1$	$1-\mu_3$
Ω^{24}	$1/(3-\mu_4-\mu_1)$	$1/\mu_4$	$1/(3-\mu_6-\mu_3)$	$2-\mu_4-\mu_2$	$1-\mu_1$	$2-\mu_6-\mu_3$
Ω^{25}	$1/(3-\mu_4-\mu_1)$	$1/(3-\mu_5-\mu_2)$	$1/\mu_3$	$2-\mu_4-\mu_2$	$2-\mu_5-\mu_1$	0
Ω^{26}	$1/(3-\mu_4-\mu_1)$	$1/(3-\mu_5-\mu_2)$	$1/\mu_6$	$2-\mu_4-\mu_2$	$2-\mu_5-\mu_1$	$1-\mu_3$
Ω^{27}	$1/(3-\mu_4-\mu_1)$	$1/(3-\mu_5-\mu_2)$	$1/(3-\mu_6-\mu_3)$	$2-\mu_4-\mu_2$	$2-\mu_5-\mu_1$	$2-\mu_6-\mu_3$

Table 4: TB affine mappings

have been computed, we are able to rewrite the weak formulation 160 into the reference domain.

Bilinear forms. Recalling the bilinear form defined in (158), we recall the definition of the conductivity tensor $\underline{\underline{K}}_r$ given in (147) (section 4.3). Due to the isotropic nature of the material, for each subdomain Ω_r^k, $1 \le k \le 27$, we need to extract three different affine terms, corresponding to the three different entries of the conductivity tensor, hence we would have $27 \times 3 = 81$ terms in our affine development. We obtain:

$$
\begin{aligned}
a(w, v; \boldsymbol{\mu}) &= \sum_{k=1}^{27} \int_{\Omega_r^k} \underline{\underline{K}}_r(\boldsymbol{\mu}) \nabla u \cdot \nabla v \, d\Omega_r^k \\
&= \int_{\Omega_r^1} \frac{\mu_2\mu_3}{\mu_1} \frac{\partial u}{\partial x_{r_1}} \frac{\partial v}{\partial x_{r_1}} + \frac{\mu_1\mu_3}{\mu_2} \frac{\partial u}{\partial x_{r_2}} \frac{\partial v}{\partial x_{r_2}} + \frac{\mu_1\mu_2}{\mu_3} \frac{\partial u}{\partial x_{r_3}} \frac{\partial v}{\partial x_{r_3}} \, d\Omega_r^1 + \dots \\
&\dots + \int_{\Omega_r^3} \left(-\frac{\mu_2(\mu_3 + \mu_6 - 3)}{\mu_1} \frac{\partial u}{\partial x_{r_1}} \frac{\partial v}{\partial x_{r_1}} + \right. \\
&\left. -\frac{\mu_1(\mu_3 + \mu_6 - 3)}{\mu_2} \frac{\partial u}{\partial x_{r_2}} \frac{\partial v}{\partial x_{r_2}} - \frac{\mu_1\mu_2}{\mu_3 + \mu_6 - 3} \frac{\partial u}{\partial x_{r_3}} \frac{\partial v}{\partial x_{r_3}} \right) d\Omega_r^3 + \dots \\
&\dots + \theta_a^{81}(\boldsymbol{\mu}) \int_{\Omega_r^{27}} \frac{\partial u}{\partial x_{r_3}} \frac{\partial v}{\partial x_{r_3}} d\Omega_r^{27}
\end{aligned} \tag{161}
$$

Linear functional. In this case the parametric linear functional (159) arises from an inhomogeneous Neumann boundary condition. This case has not been treated in section 4.3. In order to cast the integral of equation (159) into the reference domain, we proceed as follows:

$$
\begin{aligned}
f(v; \boldsymbol{\mu}) &= \int_{\Gamma_{o_6}(\boldsymbol{\mu})} q \, v \, d\Gamma_{o_6}(\boldsymbol{\mu}) \\
&= \int_{\Gamma_{r_6}} q \, v \, \underbrace{\left| \left(G^{\text{aff},k}(\boldsymbol{\mu}) \right)^{-1} \cdot \underline{e}^t \right| d\Gamma_{r_6}}_{d\Gamma_{o_6}},
\end{aligned} \tag{162}
$$

where $\underline{e}^t$ denotes the tangential unit vector and k indicates the indexes of the subdomains to which the face Γ_{r_6} belong. In particular, with regard the chosen subdomain enumeration (Figure 9), we see that Γ_{r_6} is given by

$$
\Gamma_{r_6} = \bigcup_{k=1}^{6} \Gamma_{r_6}^k . \tag{163}
$$

Therefore the linear functional can be rewritten in an affine development as:

$$\begin{aligned} f(v;\boldsymbol{\mu}) = & \int_{\Gamma^1_{r6}} \mu_2\mu_3\, q\, v\, d\Gamma^1_{r6} + \int_{\Gamma^2_{r6}} \mu_1\mu_6\, q\, v\, d\Gamma^2_{r6} + \ldots \\ & \ldots + \int_{\Gamma^6_{r6}} -\mu_5\,(\mu_3+\mu_6-3)\; q\, v\, d\Gamma^6_{r6} + \ldots \\ & \ldots + \theta^9_f(\boldsymbol{\mu}) \int_{\Gamma^9_{r6}} q\, v\; d\Gamma^9_{r6}. \end{aligned} \tag{164}$$

The affine decomposition is now clear and we have

$$a(u,v;\boldsymbol{\mu}) = \sum_1^{81} \theta^q_a(\boldsymbol{\mu}) a^q(u,v),$$

$$f(v;\boldsymbol{\mu}) = \sum_1^{9} \theta^q_f(\boldsymbol{\mu}) f^q(v),$$

where the θ-*functions* are the parameters dependent terms which appear in the bilinear form (161) and in the linear functional (164) expressed in the reference domain.
Since the geometric parameter dependence is quite involved, we will present only few results from our set of theta functions $\theta^q_a(\boldsymbol{\mu})$, $1 \le q \le 81$, $\theta^q_f(\boldsymbol{\mu})$, $1 \le q \le 9$ in Tables 5a and 5b. In the same tables we present also the definition of the $\boldsymbol{\mu}$−independent bilinear forms.

5.4 Results and visualization

We now present the results obtained by *RB* approximation for the *3D thermal block* example.
First we will give some informations about the *FE* approximation concerning the mesh, the basis function chosen.
Then we will focus on the results obtained with the *SCM* algorithm (section 3.5) for the error bounds calculations, then we will present the convergence of the *Greedy* procedure (section 3.3).
Finally we will present the output evaluation for particular combinations of the parameters, along with the *certified a-posteriori error bound*, to prove that the *RB* approximation is *reliable* and *efficient*.

***FE* discretization on the reference domain.** We represent in Figure 11 the reference domain upon we assemble *FE* components.

q	$\theta_a^q(\boldsymbol{\mu})$	$\underline{\underline{\mathbb{A}}}^q$
1	$\frac{\mu_2\mu_3}{\mu_1}$	$\int_{\Omega_r^1} \frac{\partial u}{\partial x_{r_1}} \frac{\partial v}{\partial x_{r_1}} \, d\Omega_r^1$
2	$\frac{\mu_1\mu_3}{\mu_2}$	$\int_{\Omega_r^1} \frac{\partial u}{\partial x_{r_2}} \frac{\partial v}{\partial x_{r_2}} \, d\Omega_r^1$
5	$\frac{\mu_1\mu_3}{\mu_2}$	$\int_{\Omega_r^1} \frac{\partial u}{\partial x_{r_2}} \frac{\partial v}{\partial x_{r_2}} \, d\Omega_r^1$
9	$\frac{\mu_2\mu_6}{\mu_1}$	$\int_{\Omega_r^2} \frac{\partial u}{\partial x_{r_2}} \frac{\partial v}{\partial x_{r_2}} \, d\Omega_r^2$
14	$\frac{\mu 1\mu_6}{\mu_5}$	$\int_{\Omega_r^5} \frac{\partial u}{\partial x_{r_2}} \frac{\partial v}{\partial x_{r_2}} \, d\Omega_r^5$
27	$\frac{\mu_1(\mu_2+\mu_5-3)}{\mu_3+\mu_6-3}$	$\int_{\Omega_r^9} \frac{\partial u}{\partial x_{r_3}} \frac{\partial v}{\partial x_{r_3}} \, d\Omega_r^9$
40	$\frac{\mu_5\mu_6\mu_7}{\mu_4}$	$\int_{\Omega_r^{14}} \frac{\partial u}{\partial x_{r_1}} \frac{\partial v}{\partial x_{r_1}} \, d\Omega_r^{14}$
60	$-\frac{\mu_2(\mu_1+\mu_4-3)}{\mu_6}$	$\int_{\Omega_r^{20}} \frac{\partial u}{\partial x_{r_3}} \frac{\partial v}{\partial x_{r_3}} \, d\Omega_r^{20}$
70	$\frac{\mu_5(\mu_3+\mu_6-3)}{\mu_1+\mu_4-3}$	$\int_{\Omega_r^{24}} \frac{\partial u}{\partial x_{r_1}} \frac{\partial v}{\partial x_{r_1}} \, d\Omega_r^{24}$
78	$\frac{(\mu_2+\mu_5-3)(\mu_1+\mu_4-3)}{\mu 6}$	$\int_{\Omega_r^{26}} \frac{\partial u}{\partial x_{r_3}} \frac{\partial v}{\partial x_{r_3}} \, d\Omega_r^{26}$

(a) TB $\theta_a^q(\boldsymbol{\mu})$-functions

q	$\theta_f^q(\boldsymbol{\mu})$	$\underline{\mathbb{F}}^q$
1	$\mu_1\mu_3$	$\int_{\Gamma_{r_6}^1} q\, v \, d\Gamma_{r_6}^1$
2	$\mu_1\mu_3$	$\int_{\Gamma_{r_6}^2} q\, v \, d\Gamma_{r_6}^2$
3	$\mu_2\mu_6$	$\int_{\Gamma_{r_6}^3} q\, v \, d\Gamma_{r_6}^3$
6	$-\mu_5(\mu_3+\mu_6-3)$	$\int_{\Gamma_{r_6}^6} q\, v \, d\Gamma_{r_6}^6$
9	$(\mu_2+\mu_5-3)(\mu_3+\mu_6-3)$	$\int_{\Gamma_{r_6}^9} q\, v \, d\Gamma_{r_6}^9$

(b) TB $\theta_f^q(\boldsymbol{\mu})$-functions

Table 5: *TB* $\theta(\boldsymbol{\mu})$-functions

In the figure 11 we also report the properties of the mesh and the basis functions chosen to discretize the *TB* problem.

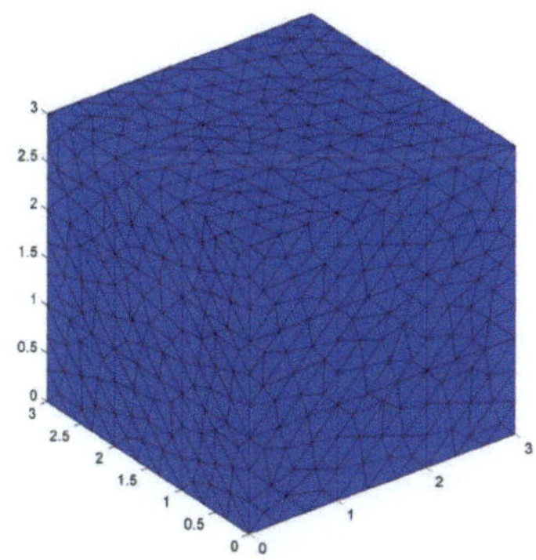

mesh vertices	1657
vertex	64
edges	432
triangle	3024
tetrahedral	7515
DOFs	1512
basis functions	$\mathbb{P}^1$

Figure 11: *TB* reference domain discretization

Matrices assembling. We now assemble the parameter independent matrices (Tables 5a and 5b) needed by the *RB* procedure.
In Figure 12 we depict a graphical view of the matrices assembling.
We consider the matrix $\underline{\underline{\mathbb{A}}}^{20}$, looking at Table 5a we note that the only subdomain that plays a rôle in the building of the parameter independent matrix is Ω_r^9, see Figure 12a. In Figure 12b we depict the matrix pattern.

***SCM* algorithm.** For the *SCM* algorithm (section 3.5) we have took a sample train Ξ_{SCM} of size $n_{\text{SCM}} = 3000$, a tolerance $\epsilon_{\text{SCM}} = 0.7$, $\mathrm{M}_\alpha = 16$, $\mathrm{M}_+ = 0$ and $|\mathcal{P}| = 200$.
In Figure 13a we show the α_{LB} (—) and the α_{UB} (—) for each element of the sample train Ξ_{SCM} of the first iteration $K = 1$, whereas in Figure 13b we depict the same quantities for the last iteration $K = K_{\max} = 4$ of the *SCM* algorithm. It is evident that the upper and lower bound for the parametric coercivity constant are converging to the exact value and restricting the possible gap between the lower and upper bound. Convergence for this problem is quite fast, see Gelsomino and Rozza (2011); Rozza et al. (2008).

***Greedy* algorithm.** We present the results for the *Greedy* algorithm (section 3.3), during the *RB* aseembling procedure.
Here, we have chosen a sample train Ξ_{train} of size is $n_{\text{train}} = 3000$, the tol-

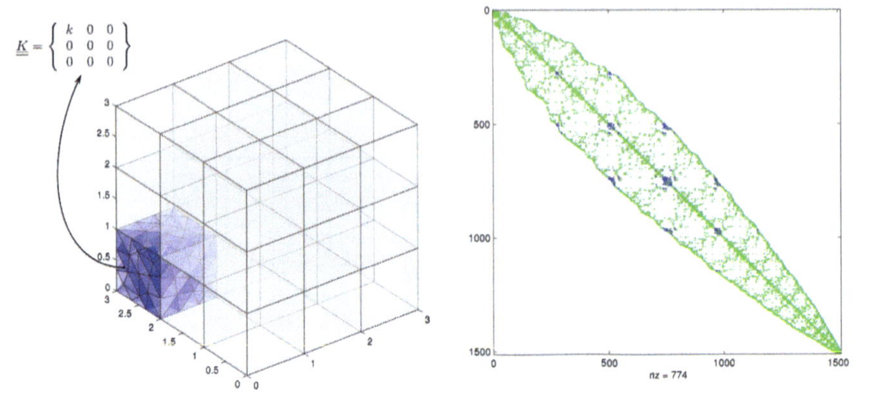

(a) subdomain Ω_r^7 (b) $\underline{\underline{\mathbb{A}}}^{20}$

Figure 12: In figure 12b we have depicted the contributes of the local $\underline{\underline{\mathbb{A}}}^{20}$ matrix (•) to the global stiffness matrix (•).

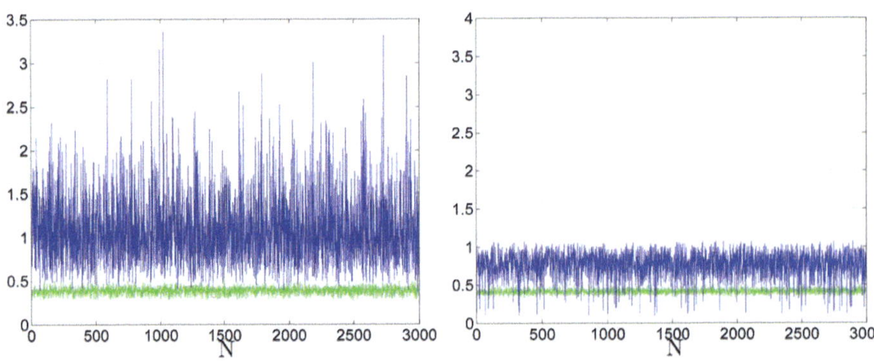

(a) *TB*: First iteration of the *SCM* (b) *TB*: Last iteration of the *SCM*

Figure 13: *TB SCM* algorithm

erance is $\epsilon_{\text{toll,min}} = 0.9 \cdot 10^{-3}$ and the maximum size of the *RB* space is taken $N_{\max} = 100$. We have chosen to minimize the absolute error bound in energy norm $\Delta_N^{en}(\boldsymbol{\mu})$ (section 3.4). In Figure 15a we have represented the error bound $\Delta_N^{en}(\boldsymbol{\mu})$ for $1 \leq N \leq N_{\max}$. We can see that the error is monotonically decreasing. Moreover, just few basis ≈ 40 (versus ≈1500 *FE DOFs*) are needed to obtain a maximum error bound $\leq 10^{-2}$ on the temperature field for all the samples in Ξ_{train}. We remark that this result holds despite a large variation of either physical (the conductivity) and ge-

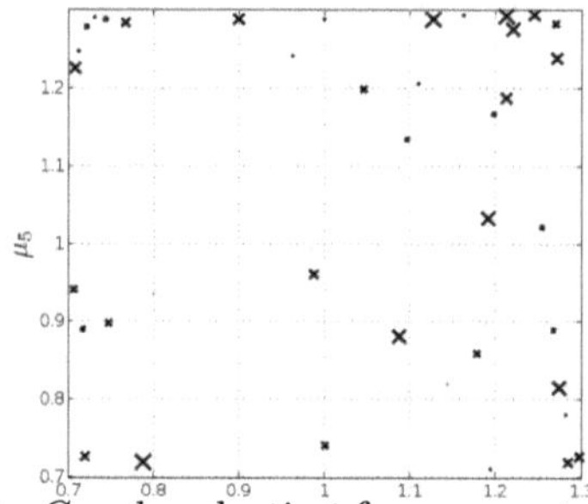

Figure 14: *Greedy* selection for parameter μ_4 and μ_5

ometrical (the dimension and the position of the central block) parameters. In Figure 15b we have depicted a subset of the parameters $\boldsymbol{\mu} = (\mu_2, \mu_5, \mu_6)$, automatically selected by the *Greedy algorithm* as representative snapshots. The error bounds help us to save also *Offline* computational cost since the evaluation of the error bounds during the *Greedy* procedure is very inexpensive. It is possible to see that the algorithm often select parameters near to the bounds (upper and lower) of the parameters domain. In fact the more the parameters are chosen distant from the "center" of the set, the most the reference domain is deformed by means of the geometrical parameters. These phenomena will perforce increase the error bound, therefore the Greedy algorithm will preferably choose this outer parameters, being based on the *worst case scenario*.
This aspect can be better seen, looking at the Figure 14, we can see that the geometrical parameters chosen are always in the outer part of the domain, where it is evident a *clustering phenomena*. In this figure, the dimension of the markers are proportional to the maximum error bound at the K-th Greedy iteration.

Output. Since the output is the average of the temperature on the face Γ_6 of the domain, then we are dealing with a compliant case (see section 2.3). In fact we have that:

$$s_N(\boldsymbol{\mu}) = f(u_N; \boldsymbol{\mu}). \tag{165}$$

Since we have 7 parameters, we decided to fix some parameters and add relationship between others to obtain a graphical visualization of the output. In particular we have chosen to vary the parameters μ_5 and μ_4, i.e. the x and y dimension of the inclusion. In addition we have introduced the following

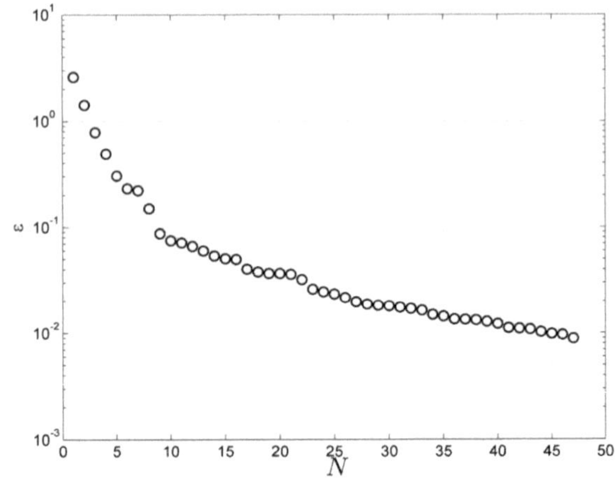

(a) Error bound $\Delta_N^{en}(\boldsymbol{\mu})$

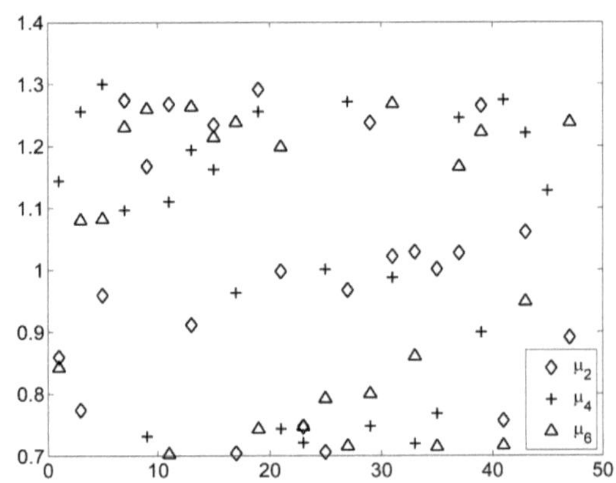

(b) Parameters distribution

Figure 15: *TB* Greedy results

relationships:

$$\mu_1 = \frac{3-\mu_4}{2};$$
$$\mu_2 = \frac{3-\mu_5}{2};$$
$$\mu_3 = 1;$$
$$\mu_6 = 1;$$
$$\mu_7 = 1.$$

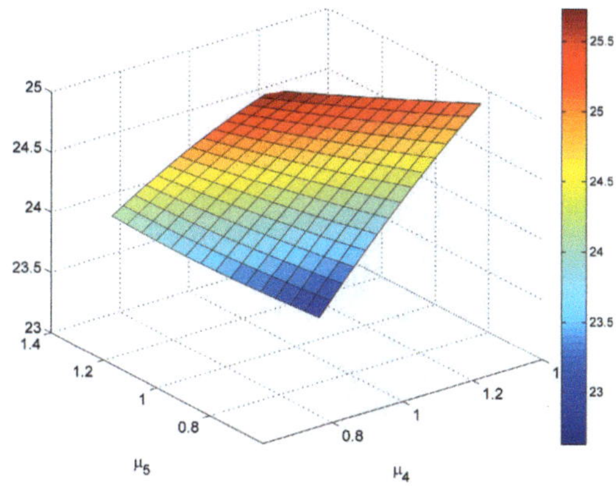

(a) *TB* output $s(\boldsymbol{\mu})$

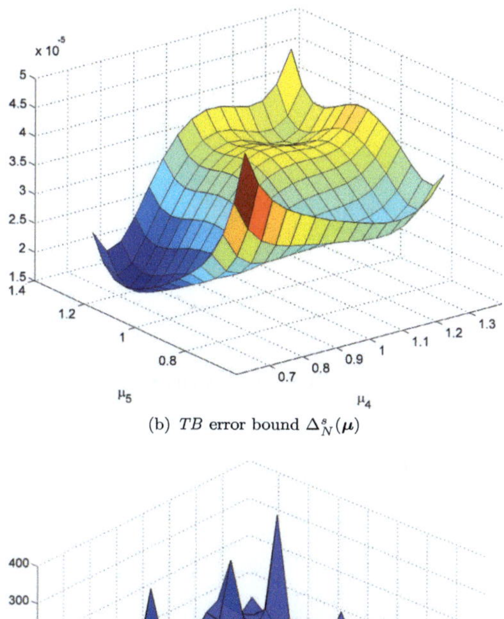

(b) *TB* error bound $\Delta_N^s(\boldsymbol{\mu})$

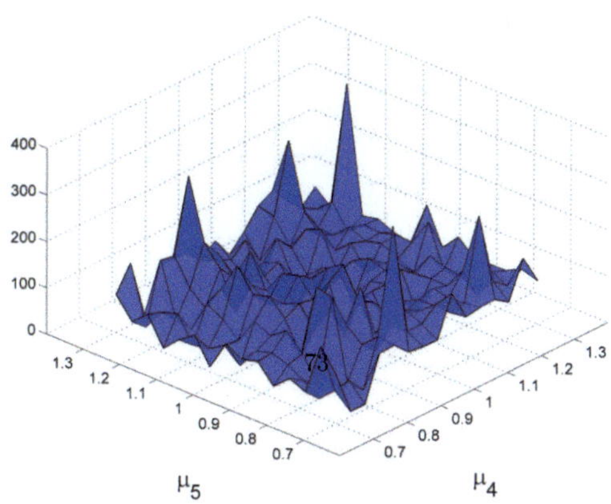

(c) $\frac{t^s_{\mathrm{FE}}(\boldsymbol{\mu})}{t^s_{\mathrm{RB}}(\boldsymbol{\mu})}$

Figure 16: *TB* output results

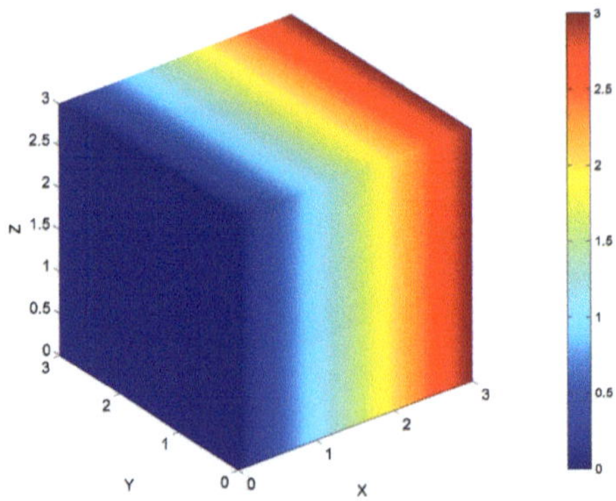

(a) Solution field: $\Delta_N^{\mathrm{en}}(\boldsymbol{\mu}) \leq 2.3 \cdot 10^{-3}$

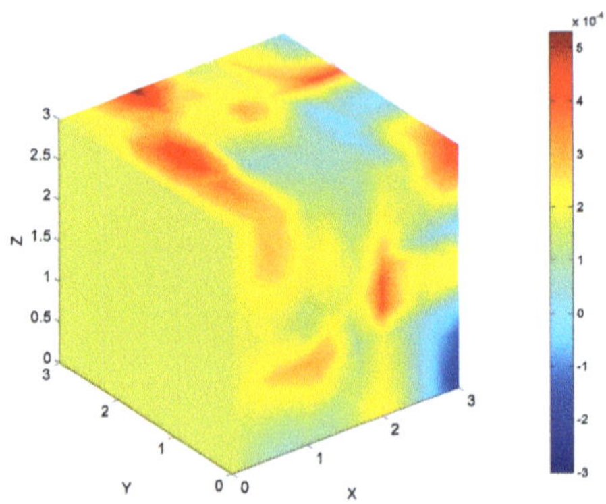

(b) Pointwise error: $u_N(\boldsymbol{\mu}) - u^{\mathcal{N}}(\boldsymbol{\mu})$

Figure 17: Example of representative solution for the *TB* problem and pointwise error for $\boldsymbol{\mu} = \{1, 1, 1, 1, 1, 1, 1\}$

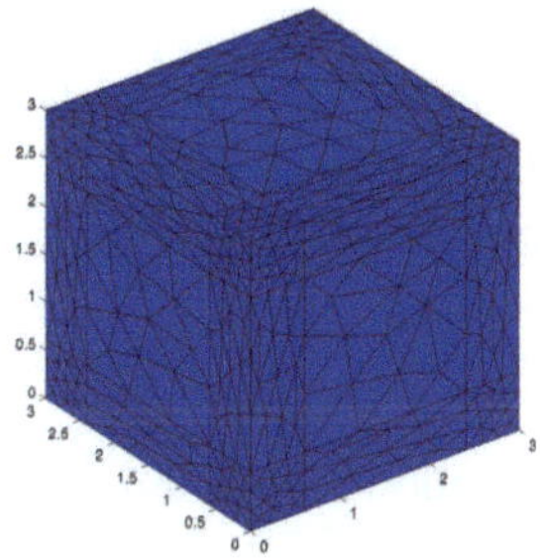

(a) Deformed domain

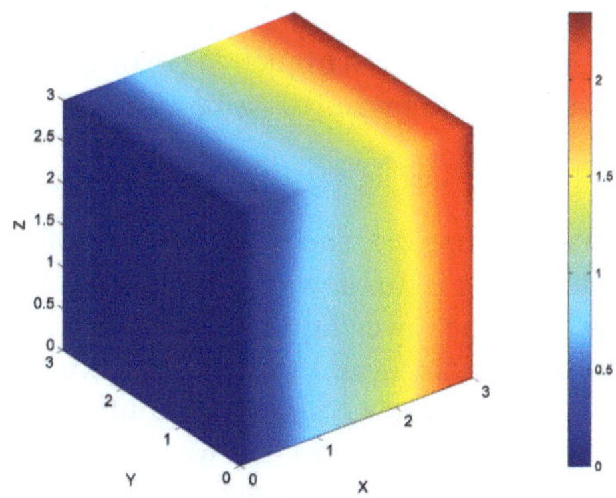

(b) Error bound: $\Delta_N^{\text{en}}(\boldsymbol{\mu}) \leq 6.3 \cdot 10^{-3}$

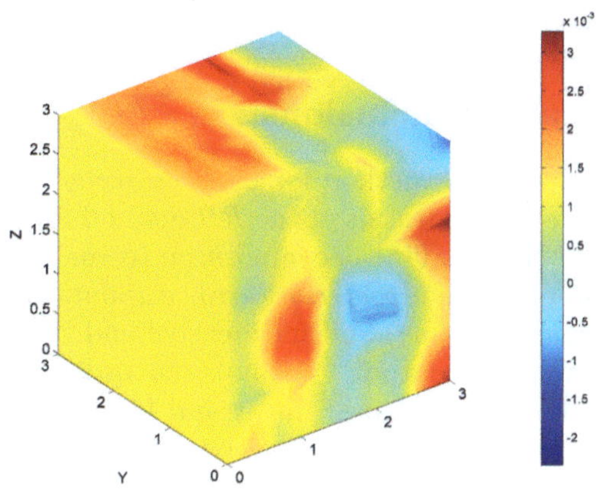

(c) Pointwise error: $u_N(\boldsymbol{\mu}) - u^{\mathcal{N}}(\boldsymbol{\mu})$

Figure 18: Example of representative solution and pointwise error for $\boldsymbol{\mu} = \{0.7, 0.7, 0.7, 1.3, 1.3, 1.3, 0.05\}$

In Figure 16a we depict the tenmperature output obtained with the *RB* method, in Figure 16b we depict the *error bound* $\Delta_N^s(\boldsymbol{\mu})$ on the output between the *FE* and the *RB* method.
We can see that the output estimated error is under $8 \cdot 10^{-5}$, hence the error on the output is effectively bounded by the square of the error on the solution field (we recall that in the Greedy we have set $\Delta_N^{\text{en}}(\boldsymbol{\mu}) \leq \epsilon_{\text{toll,min}} = 0.9 \cdot 10^{-3}$, $\forall \boldsymbol{\mu} \in \Xi_{\text{train}}$).
We remark that this result follows from our assumption of compliance, enabling the so called *square effect* (57).
In Figure 16c we have depicted the ratio between the computational time needed to evaluate the output in the *FE* (denoted with $t^s_{\text{FE}}(\boldsymbol{\mu})$) and in the *RB* case ($t^s_{\text{RB}}(\boldsymbol{\mu})$) for a large test sample. We can see that the *RB* method provides a computational time saving of *two order of magnitude* with respect to the ordinary *FE* method.

Visualization. We now report the visualization of some representative *RB* solutions. On the upper figures, we show the solution for different value of the parameters $\boldsymbol{\mu}$. On the lower figures, we represent the pointwise error between the *RB* approximations and the *FE* solution.
In the first example, Figure 17, we show the solution on the reference domain. In the second example, Figure 18, we show the solution field after selecting a generic combination of parameters in the parameter domain $\mathcal{D}$.
In the first case, thanks to the absence of geometrical distortion, we obtain a *smaller error bound* on the solution.

6 Summary and conclusions

We have presented the *RB* method for *coercive scalar* problems. First, we have introduced the fundamentals of the *RB* method for *parametrized coercive elliptic PDEs*: (i) the affine decomposition to enable an Offline/Online splitting procedure, (ii) the *a-posteriori error estimates* to efficiently create the *RB* Greedy space (Offline) and inexpensive and rigorous error bounds for the *RB* solution and output (Online). In order to obtain an affine representation of the parametrized linear and bilinear forms we exploited an *affine geometry precondition.* We presented the applications of the *RB* method to a *steady thermal conductivity problem* in heat transfer.
We obtained a good and rapid convergence of either the *SCM* and the Greedy algorithm. Hence we experimentally proved that the *RB* method is very well suited to efficiently approximate also 3D problems with a rather involved parameter dependence, either physical and geometrical.
The Offline stage is quite expensive in the 3D context, nonetheless the very

inexpensive and rigorous Online stage renders invaluable the worth of the *RB* method in many engineering field of interests: optimization, control, sensitivity analysis and real-time context.
In fact by the *RB* method we obtained (Online) a computational saving of at least two order of magnitude with respect to the *FE* approximation in the thermal application, corroborating other results (Gelsomino and Rozza, 2011) dealing with 3D coercive problems.

Different branches of the research field related to the *RB* method are dealing nowadays with a plenty of different problems and different contexts: the study of potential flows (Rozza, 2011), thermal problems, (Rozza et al., 2009b,a), hemodynamics and biomedical devices optimization (Manzoni et al., 2012), the study of nonlinear equations such as the Navier-Stokes problem, (Deparis and Rozza, 2009), the development of *RB* approximation in parabolic (Nguyen et al., 2010) and hyperbolic equations (Nguyen et al., 2009). Unfortunately, the affine parametrization of a geometry, is not enough *flexible* for some purposes, e.g. the shape-optimization of a vessel in a hemodynamic context or the design of an *air intake* in an aerodynamic context. Hence in the few last years, new techniques based on the so-called *free form deformation* has been re-invented in cooperation with the *RB* method (Manzoni et al., 2012; Lassila and Rozza, 2010; Rozza and Manzoni, 2010) in the shape optimization field.
A possible and remarkable upgrade of this work would be to enrich the 3D geometrical parametrization with *free-shape* techniques to improve the power of reduced order methods in the engineering context.

Bibliography

B.O. Almroth, P. Stern, and F.A. Brogan. Automatic choice of global shape functions in structural analysis. *AIAA Journal*, 16(7):525–528, 1978.

V.S. Arpaci. *Conduction Heat Transfer*. Addison-Wesley, 1966.

J.A. Atwell and B.B. King. Proper Orthogonal Decomposition for Reduced Basis Feedback Controllers for Parabolic Equations. *Mathematical and computer modeling*, 33(1-3):1–19, 1999.

T. Bui-Thanh, K. Willcox, and O. Ghattas. Model Reduction for Large-Scale Systems with High-Dimensional Parametric Input Space. *Proceedings of the 48th AIAA/ASME/ASCE/AHS/ASC structures, structural dynamics and material conference*, 30(6):3270–3288, 2008.

S. Deparis and G. Rozza. Reduced Basis Method for Multi-Parameter dependent Steady Navier-Stokes Equations: Applications to Natural Convection in a cavity. *Journal of Computational Physics*, 228:4359–4378, 2009.

R. Fox and H. Miura. An approximate analysis technique for design calculations. *AIAA Journal*, 9(1):177–179, 1971.

F. Gelsomino and G. Rozza. Comparison and combination of reduced order modeling techniques in 3d parametrized heat transfer problems. *Mathematical and Computer Modelling of Dynamical Systems*, 17(4):373–391, 2011.

M.A. Grepl, Y. Maday, N.C. Nguyen, and A.T. Patera. Efficient Reduced Basis Treatment of Nonaffine and Nonlinear Partial Differential Equations. *European Series in Applied and Industrial Mathematics: Mathematical Modelling and Numerical Analysis*, 41(3):575–605, 2007.

B. Haasdonk and M. Ohlberger. Reduced Basis method for Finite Volume Approximations of Parametrized Linear Evolution Equations. *Mathematical Modelling and Numerical Analysis*, 42(2):277–302, 2008.

D.B.P. Huynh, G. Rozza, S. Sen, and A.T. Patera. A Successive Constraint Linear Optimization Method for Lower Bounds of Parametric Coercivity and Inf–Sup Stability Constants. *Comptes Rendus de l'Academie des Sciences*, 345(8):473–478, 2007.

K. Ito and S.S. Ravindran. Reduced Basis method for Optimal Control of Unsteady Viscous Flows. *International Journal of Computational Fluid Dynamics*, 15:97–113, 2001.

K. Ito and S.S. Ravindran. A Reduced Order Method For Control Problems Governed by PDEs. volume 126 of *International Series of Numerical Mathematics*, pages 153–168. Birkhauser, Boston, 1998.

M. Jabbar and A. Azeman. Fast optimization of Electromagnetic-Problems: the Reduced Basis Finite Element Approach. *IEEE Transactions on Magnetics*, 40(4):2161–2163, 2004.

K. Kunisch and S. Volkwein. Galerkin Proper Orthogonal Decomposition Methods for a General Equation in Fluid Dynamics. *SIAM Journal on Numerical Analysis*, 40(2):492–515, 2003.

T. Lassila and G. Rozza. Parametric Free-Form Shape Design with PDE Models and Reduced Basis Method. *Computer Methods in Applied Mechanics and Engineering*, 199:1583–1592, 2010.

Barrault M., Maday Y., Nguyen N.C., and Patera A.T. An Empirical Interpolation Method: Application to Efficient Reduced Basis discretization of Partial Differential Equations. *Comptes Rendus Mathematiques*, 339 (9):667–672, 2004.

A. Manzoni, A. Quarteroni, and G. Rozza. Shape Optimization in Cardiovascular Geometries using Reduced Basis Method and Free-Form Deformations. *International Journal for Numerical Methods in Fluids*, 70(5): 646–670, 2012.

C.D. Meyer. *Matrix analysis and applied linear algebra.* SIAM, Philadelphia, Colorado State University, 1st edition, 2000.

D.A. Nagy. Modal representation of geometrically nonlinear behaviour by the finite element method. *Computers & Structures*, 10(4):683–688, 1979.

N.C. Nguyen, K. Veroy, and A.T. Patera. *Certified Real-Time Solution of Parametrized Partial Differential Equations.* Yip S(ed) Handbook of materials modeling. Springer, Berlin, 2005.

N.C. Nguyen, G. Rozza, and A.T. Patera. Reduced Basis Approximation and A-Posteriori Error Estimation for Time Dependent Viscous Burgers Equation. *Calcolo*, 46(3):157–187, 2009.

N.C. Nguyen, G. Rozza, D.B.P. Huynh, and A.T. Patera. Reduced Basis Approximation and A-Posteriori Error Estimation for Parametrized Parabolic PDEs; Application to Real-Time Bayesian Parameter Estimation. In *Computational Methods for Large Scale Inverse Problems and Uncertainty Quantification*, pages 151–173. John Wiley & Sons, UK, 2010.

A.K. Noor. On making large nonlinear problems small. *Computer Methods in Applied Mechanics and Engineering*, 34(2):955–985, 1982.

A.K. Noor. Recent advances in reduction methods for nonlinear problems. *Computers & Structures*, 13(1-3):31–44, 1978.

A.K. Noor and J.M. Peters. Reduced Basis Technique for Nonlinear Analysis of structures. *Computers & Structures*, 18(4):455–462, 1980.

A.T. Patera and E.M. Ronquist. On the error behavior of the Reduced Basis Technique for Nonlinear Finite Element Approximations and A-Posteriori Error estimation for a Boltzmann model. *Computer Methods in Applied Mechanics and Engineering*, 196:2925–2942, 2007.

A.T. Patera and G. Rozza. *Reduced Basis Approximation and A-Posteriori Error Estimation for Parametrized Partial Differential Equations.* MIT Pappalardo Graduate Monographs in Mechanical Engineering. ©MIT, Massachusetts Institute of Technology, Cambridge, MA, USA, 2007.

J.S. Peterson. The Reduced Basis method for incompressible viscous flow calculations. *SIAM Journal on Scientific Computing*, 10(4):777–786, 1989.

T.A. Porsching. Estimation of the Error in the Reduced Basis Method Solution of Nonlinear Equations. *Mathematics of Computation*, 45(172): 487–496, 1985.

T.A. Porsching and M.L. Lee. The Reduced Basis method for Initial Value Problems. *SIAM Journal on Numerical Analysis*, 24(6):1277–1287, 1987.

C. Prud'homme, D. Rovas, K. Veroy, Y. Maday, A. Patera, and G. Turinici. Reliable real-time solution of parametrized partial differential equations: Reduced Basis output bounds methods. *Journal of Fluid Engineering*, 124(1):70–80, 2002a.

C. Prud'homme, D.V. Rovas, K. Veroy, and A.T. Patera. A mathematical and Computational Framework for Reliable Real-Time Solution of Parametrized Partial Differential Equations. *Mathematical Modeling and Numerical Analysis*, 36(5):747–771, 2002b.

A. Quarteroni. *Numerical Models for Differential Problems*, volume 8 of *MS&A*. Springer, 2013.

A. Quarteroni and A. Valli. *Numerical Approximation of Partial Differential Equations*. Springer series on Computational Mathematics. Springer, 1997.

A. Quarteroni, R. Sacco, and F. Saleri. *Numerical Mathematics*. Texts in Applied Mathematics. Springer, New York, 2000.

S.S. Ravindran. Adaptive reduced order controllers for a thermal flow system using Proper Orthogonal Decomposition. *SIAM Journal on Scientific Computing*, 23(3):1924–1942, 2002.

D. Rovas. *Reduced Basis Output Bound Methods for Parametrized Partial Differential Equations*. PhD thesis, Massachusetts Institute of Technology, Cambridge, MA, USA, 2003.

G. Rozza. Reduced basis approximation and error bounds for potential flows in parametrized geometries. *Communications in Computational Physics*, 9(1):1–48, 2011.

G. Rozza. Reduced Basis for Stokes equations in domains with non-affine parametric dependence. *Computing and Visualization in Science*, 12(1): 23–35, 2009a.

G. Rozza. An Introduction to Reduced Basis Method for Parametrized PDEs. In *In Applied and Industrial Mathematics in Italy, Vol. III, Series on Advances in Mathematics for Applied Sciences*, pages 508–519. Societá Italiana di Matematica Applicata e Industriale, 2009b.

G. Rozza and A. Manzoni. Model Order Reduction by Geometrical Parametrization for Shape Optimization in Computational Fluid Dynamics. In *Proceedings of ECCOMAS CFD Conference, J.Pereira and A. Sequeira Editors*, Lisbon, Portugal, June, 2010.

G. Rozza and K. Veroy. On the Stability of the Reduced Basis method for Stokes Equations in parametrized domains. *Computer Methods in Applied Mechanics and Engineering*, 196:1244–1260, 2006.

G. Rozza, D.B.P. Huynh, and A.T. Patera. Reduced Basis approximation and A-Posteriori Error Estimation for Affinely Parametrized Elliptic Coercive Partial Differential Equations. *Archives of Computational Methods in Engineering*, 15:229–275, 2008.

G. Rozza, D.B.P. Huynh, N.C. Nguyen, and A.T. Patera. Real-Time Reliable Simulation of Heat Transfer Phenomena. In *Heat Transfer Summer Conference Proceedings, HT-2009-88212*, S. Francisco, CA, US, July 2009a. American Society of Mechanical Engineers.

G. Rozza, N.C. Nguyen, A.T. Patera, and S. Deparis. Reduced Basis Methods and A-Posteriori Error Estimators for Heat Transfer Problems. In *Heat Transfer Summer Conference Proceedings, HT-2009-88211*, S. Francisco, CA, US, July 2009b. American Society of Mechanical Engineers.

P. Walker. *Chambers Dictionary of Materials Science and Technology.* Chambers Publishing, 1st edition, 1993.

K. Willcox and J. Peraire. Balanced Model Reduction via the Proper Orthogonal Decomposition. *AIAA Journal*, 40(11):2323–2330, 2002.

K. Yosida. *Functional Analysis.* New York: Springer-Verlag, 1971.

MIX
Papier aus verantwortungsvollen Quellen
Paper from responsible sources
FSC® C105338

If you have any concerns about our products,
you can contact us on
ProductSafety@springernature.com

In case Publisher is established outside the EU,
the EU authorized representative is:
Springer Nature Customer Service Center GmbH
Europaplatz 3, 69115 Heidelberg, Germany

Printed by Libri Plureos GmbH
in Hamburg, Germany